T0139956

# Elements of Quantum Computing

Seiki Akama

# Elements of Quantum Computing

## History, Theories and Engineering Applications

 Springer

Seiki Akama
Kawasaki
Japan

ISBN 978-3-319-36213-7          ISBN 978-3-319-08284-4     (eBook)
DOI 10.1007/978-3-319-08284-4
Springer Cham Heidelberg New York Dordrecht London

Printed on acid-free paper

Springer is part of Springer Science+Business Media (www.springer.com)

# Preface

Now, computer technology is rapidly developed. It appears that the computer world is completely established. Certainly, the fact may be true according to current thinking but there seems to be a room for further developments. This is because we have some possibilities to realize a computer from a different point of view.

One of such possibilities is the so-called *quantum computer*. It is a computer based on quantum mechanics, and it can do super speed computation. It is well known that theoretical possibilities of quantum computer have been argued since the beginning of the 1980's.

There has been a growing interest in the 1990's, and some quantum computers at the experimental level were recently implemented. However, most people do not appear to be familiar with the field of quantum computer.

It is also true for experts working on computer science. The reason is that we cannot understand quantum computers without the background of quantum mechanics. It can be also pointed out that the foundations for quantum computer are not always obvious.

We believe that one must study the basics of quantum computing systematically. Unfortunately, it is difficult to read most papers, including survey papers, on quantum computing. If there is a readable text book on the subject, many people will be able to understand (or study) quantum computing.

The purpose of this book is to provide a quick introduction to quantum computing for the readers who have no backgrounds of both theory of computation and quantum mechanics.

But to understand quantum computing, we need the knowledge of both computer science and quantum mechanics. One cannot easily learn relevant materials in both fields. Thus, we give an intuitive explanation of the material, and then add theoretical justification.

The structure of the book is as follows.

Chapter 1 gives a quick review of quantum computing and quantum mechanics. First, we outline a quantum computer to understand quantum computing. Second, we briefly explain the ideas of quantum mechanics. We also describe the history of both quantum computing and quantum mechanics.

Chapter 2 surveys current computer models including Neumann-type computer, Turing Machine and Boolean algebra. Before looking at quantum computers, we have to understand these models properly.

Chapter 3 introduces quantum mechanics. After looking at the basic ideas of quantum mechanics, wave mechanics, matrix mechanics, and the uncertainty principle are explained. Since the materials in this chapter are mathematically advanced, the readers who are not interested in them can skip and proceed to Chapter 4.

Chapter 4 explains quantum computer in some details. After the historical survey of the origins of quantum computers, we introduce the quantum Turing Machine. Qubit, quantum gates and Shor's algorithm are also discussed.

Chapter 5 outlines the representative applications of quantum computing. We take up quantum codes, quantum communications, quantum teleportation and quantum programming.

Chapter 6 concludes the book with discussing the future of quantum computing. In particular, implementations and important problems of quantum computing are discussed.

We do not assume that the reader has any special knowledge about theory of computation and quantum mechanics. We hope that the book is of help for the readers who want to understand quantum computing quickly.

Seiki Akama
May 2014

# Contents

# Chapter 1
# Introduction

**Abstract.** Chapter 1 gives a quick review of quantum computing and quantum mechanics. First, we outline a quantum computer to understand quantum computing. Second, we briefly explain the ideas of quantum mechanics. We also describe the history of both quantum computing and quantum mechanics.

## 1.1 What Is Quantum Computing?

*Quantum computer* is a computer based on a computational model which uses *quantum mechanics*. As you know, quantum mechanics is a subfield of physics which studies phenomena at the micro level. The area related to quantum computers is generally called the *quantum computing*, which is regarded as a promising field in computer science.

### 1.1.1 Intuitive Definition of Quantum Computers

Quantum computers do computation based on the theory of quantum mechanics, whose computation is performed very different from traditional ones, e.g., Neumann-type computer. Perhaps, the reader may not be able to understand the description. But, in fact, we now have such computers, although they are experimentally built.

It is here necessary for the reader to have mathematical background in order to grasp the basics of quantum computers. We believe that it is valid to define quantum computers in this way, and we will defer giving formal exposition. However, it does not seem to be a good idea for beginners.

Because we cannot easily learn an abstract theory in general, it is helpful to obtain intuitive image of it. This can be applied to the study of quantum computers. Here, one would like to imagine the following picture.

Namely, quantum computers represent information by making use of phenomena in atoms at the micro level and employ quantum mechanical systems with some properties to do computations.

Consequently, quantum computers have a hardware and a computational mechanism which are radically different from those in previous computers. This is one of the most important points in understanding quantum computers.

There are some possible descriptions of the intuitive picture of quantum computers. We could here give one of the images of actual computation in quantum computers as follows:

> When certain computation is performed in quantum computers, *similar computation* is *simultaneously* performed in *other world* which is connected with the *actual world*. The result is obtained by a probability if we *try to see* the result of computation, and we can obtain the probability that the result holds.

The principle of quantum computers is summarized as above, where the italicized parts express the important keywords. The reader will realize why these terms are written in italics, if the reader reads over the book.

In the sequel, we give the exposition by using technical terms. The image presented here is useful for the reader to understand the materials of the book. Of course, it is better to mathematically understand the materials.

Quantum computer is a computer whose computation is performed *quantum mechanically*. It has a new type of hardware with the computational mechanism based on quantum mechanics. In other words, both the hardware and software should be constructed based on the principles of quantum mechanics.

The basic unit of *information* in quantum computers is called a *qubit*. It is based on *superposition states* in which both the states 0 and 1 are overlapped. Current computers represent information by means of a *bit* which can have only one value of 0 and 1. On the other hand, quantum computers use qubits which can represent more information than in bits.

Thus, if a number of qubits are given and parallelly processed, then quantum computers enable super speed computation. As a result, quantum computers can perform the computation of some important problems whose solutions were regarded impossible or intractable in traditional computers.

It is generally thought that the advent of quantum computers can change the world of computers and our life. Some people showed that quantum computers can be used to compute the answers to certain difficult problems much faster than any classical computers. It is in fact a very interesting fact. Now, work for implementing quantum computers, though at the experimental level, is in progress all over the world.

### 1.1.2  History of Quantum Computers

Here, we briefly look at the history of quantum computers. So far, many people have suggested the ideas of quantum computers and formalized their models. To survey the history is of interest.

## Benioff

In 1980, Benioff pointed out a possibility of constructing a computer based on quantum mechanics; see Benioff [6] for details. He proposed a quantum mechanical model of computers within the framework of *Turing Machine*, which is a standard model of computation.

Benioff's starting point was to make smaller logic circuits. Benioff showed that an atomic-scale circuit could be built. Benioff's model can be interpreted as a computer model which satisfies the laws of quantum mechanics.

Benioff suggested to use the different spins of elementary particles to represent two binary digits. Practically, in his model, the computation is quantum-mechanically performed without energy consumption.

However, his model is essentially equivalent to Turing Machine which is one of the traditional computational models. As a consequence, it cannot exceed the computational power of a classical computer. Thus, his model cannot be interpreted as a *new* model for quantum computers.

## Feynman

In 1982, Feynman, who received the Nobel Prize in Physics in 1965, discussed a quantum computer which can simulate physics in Feynman [36]. Although Feynman's contributions to quantum computers are less known, his paper is undoubtedly one of the origins of quantum computers.

Feynman pointed out that some kind of quantum mechanical phenomena cannot be effectively simulated by traditional classical computers and claimed that "computation" for simulating physics can be performed more effectively by using quantum mechanical phenomena.

According to Feynman, quantum computers should have a computational mechanism which obeys the laws of quantum mechanics. Additionally, Feynman introduced some examples of physical phenomena which can be interpreted as a computational model for quantum computers. We will give a detailed account on Feynman's ideas in Chapter 4.

## Deutsch

In 1985, English physicist Deutsch proposed a computational model called the *quantum computer* which reformulates a Turing Machine and has the computational power equivalent to *quantum Turing Machine*; see Deutsch [24].

Deutsch's "quantum computer", as reviewed in Chapter 4, is the first computational model for quantum computing. Because his theory is more concrete than Feynman's and compatible to the tradition of computer science, it should be considered as another origin of the research on quantum computers.

Deutsch also showed a theory of *quantum gate*, which is another model of quantum computers, in 1988; see Deutsch [25]. A quantum gate is a logical gate for quantum computers. Deutsch gave universal quantum gate, and this increased a possibility of realizing quantum computers.

**Bernstein and Vazirani**

In 1993, Bernstein and Vazirani proposed a *universal quantum Turing Machine* which generalizes Deutsch's quantum Turing Machine; see Bernstein and Vazirani [11]. They seemed to give a theory of universal quantum Turing Machine in the traditional form.

They also proved that both (classical) Turing Machine and quantum Turing Machine have the equivalent computational power. Their research became a first step to clarifying the computational aspects of quantum computers like computability and complexity. In this regard, they first provided contributions to quantum complexity theory.

**Shor**

In 1994, Shor demonstrated a polynomial-time quantum algorithm for prime factorization and discrete logarithm; see Shor [72],[73]. Shor's result was astonishing and later attracted our interest of quantum computers.

A quantum algorithm for prime factorization is called *Shor's algorithm*. It is observed the algorithm justifies a strong possibility of quantum computers. This is because prime factorization is known one of the intractable problems which is the underlying base of *RSA code*. It is worth noting that Shor's algorithm showed a quantum algorithm for the concrete problem.

Shor's result reveals that there is a possibility of breaking the safety of RSA coding if a quantum computer is implemented and prime factorization is fastly performed by it. The result was surprising for computer science. Consequently, the importance of quantum computers was recognized.

**Grover**

In 1996, Grover proposed quantum search algorithm in Grover [44]. Grover showed the quantum algorithm which can search data in unstructured database in the order of $\sqrt{n}$. The algorithm is called the *Grover's algorithm*.

**Omer**

In 1998, Omer implemented a programming language for quantum computers called *QCL* at Technical University of Vienna; see Omer [64]. QCL is a procedural language like the C language and can be used for implementation and simulation for various quantum algorithms. By the language, it became possible to study quantum commuting in the software level.

By the late 1990's, most foundational research on quantum computers has been done, and some research on implementation has also started. But many people thought the problem of how the hardware of quantum computers should be constructed is extremely difficult.

**Gershenfeld and Chuang**

There are several types of hardware of quantum computers which are different from the one of traditional computers.

In 1998, Gershenfeld and Chuang at MIT developed a 2-qubits quantum computer based on *NMR* (Nuclear Magnetic Resonance); see Gershenfeld and Chuang [41]. This is the first actual implementation of a quantum computer.

In 2001, IBM succeeded in the development of 7-qubits NMR quantum computer and implemented Shor's Algorithm, thus showing the effectiveness of NMR for the implementation for quantum computers; see Vandersypen et al. [80].

There are also other approaches to the hardware of quantum computers. They include trapped ion, quantum dots, and Josephson junction. Currently, it is possible to develop quantum computers with more than 10-qubits and further investigations are expected. We note that fundamental theories for applications of quantum computing are being established.

## 1.2 What Is Quantum Mechanics?

*Quantum mechanics* (QM) is the theoretical basis of quantum computing. Here, we only introduce its basic ideas and history. The details of mathematical aspects of quantum mechanics will be given in Chapter 3.

### 1.2.1 Classical Physics

There are various *matters* in our world. Matter is regarded as anything that has both mass and volume. *Physics* studies the nature of matter mathematically. As you know, the *atom* is a basic unit of matter.

An *elementary particle* is a particle that has substructures. An atom consists of several elementary particles. There are many elementary particles such as photon, electron, proton and neutron.

The proton and neutron are composed of elementary particles which are called *quarks*. A quark is also an elementary particle which is regarded elementary and indivisible.

Matters are the existence of the macro world, and we can see them. But atoms and elementary particles are the existence of the micro world . By the 19th century, physics dealt with matters. But it also considered atoms and elementary particles since the 20th century.

Physics of matters in the macro level is considerably different from that of matters in the micro level, i.e., atoms and elementary particles. In this regard, the physics of the former is generally referred to as *classical physics*, and the latter as *quantum theory (physics)*.

Classical physics was systematized by Newtonian mechanics and electromagnetism. Newton mathematically expressed the properties of motion of objects as the so-called *Newton's laws*. Maxwell provided foundations for electromagnetism by means of the so-called *Maxwell's equations*. It is common knowledge that classical physics was established by the end of the 19th century.

## 1.2.2 Quantum Theory

Quantum theory, also called the *modern physics*, includes quantum mechanics. And the theory before quantum mechanics is often called the *old quantum theory*. Usually, quantum theory is synonymous to quantum physics. Note also that the Theory of Relativity belongs to modern physics.

**Newton**

In the 17th century, Newton established the (classical) *mechanics*, which can formalize the movements of objects. In 1687, Newton stated three laws of motion:

---

(1) Every body continues in a state of rest, or of uniform motion in a straight line, unless it is compelled to change that state by forces impressed upon it.

(2) Any change of motion is proportional to the force that acts, and it is made in the direction of the straight line in which that force is acting.

(3) To every action there is always an equal and opposite reaction; or, the mutual actions of two bodies upon each other are always equal and act in opposite directions.

---

Newton's first law corresponds to Galileo's law of inertia. The second law can be formalized as the equation $F = ma$, where $F$ is a force, $m$ is a mass and $a$ is an acceleration. The third law concerns the relation of an action and its reaction.

Newton's work can be regarded as the first mathematical formulation of mechanics which consists of the above simple principles. Classical mechanics is the theoretical basis for describing the motion of macroscopic objects.

**Isaac Newton (England: 1642-1727)**

Isaac Newton was born in Woolsthorpe, England on 1642. He entered Cambridge University in 1661. He contributed to several fields in natural science, in particular, physics and mathematics. He formally described the classical mechanics including the law of universal gravitation and the three laws of motion, in *Philosophiae Naturalis Principia Mathematica* published in 1687. He also founded the theory of calculus, which was disputed with Leibniz. Newton died in Kensington, England on 1727.

**Einstein**

In 1905, Einstein proposed *the special theory of relativity*, and in 1915, *the general theory of relativity*, which supplement the classical mechanics.

The first showed that Newton's three laws of motion can be only approximately correct, breaking down when velocities approached that of light. The second showed that Newton's law of gravitation can be also only approximately correct, breaking down in the presence of very strong gravitational fields. By the work of Newton and Einstein, the classical mechanics is considered to be completed.

**Albert Einstein (Germany: 1879-1955)**

Albert Einstein was born in Ulm, Germany on 1879. In 1896, he entered the Swiss Federal Polytechnic School in Zurich. In 1905, he received doctor's degree. In 1911, he became Professor of Theoretical Physics at Prague. He contributed to the modern physics. In particular, the special theory of relativity and the general theory of relativity. He received the Nobel Prize in Physics in 1921 for his study of photoelectric effect rather than the theory of relativity. Einstein died in Princeton on 1955.

Entering in the 20th century, the first task of the quantum theory was to clarify the true character of a light. Namely, it is the dispute whether a light is a wave or a particle. By the time, a light has been considered as a wave, because it leads to the phenomena called *interference*, which is the interaction of two or more waves passing the same point, producing one combined wave.

In the early 19th century, English scientist Young performed the well known double-slit experiment which demonstrated that a beam of light, when split into two beans and then recombined, will show interference that can only be explained by assuming that light is a wave.

**Planck**

In 1900, Planck discovered the so-called Planck's law which says that the energy of electromagnetic radiation is confined to indivisible packets (quanta), each of which has an energy equal to the product of the Planck constant and the frequency of the radiation, and claimed that a light is a particle. His claim is called the *quantum hypothesis*. Plank's work initiated the study of quantum theory.

**Max Planck (Germany: 1858-1947)**

Max Planck was born in Kiel, Germany in 1858. Planck entered the University of Münich in 1874, and received the doctoral degree in 1879. In 1900, Planck introduced the notion of a quantum to motivate the view that energy is transmitted in particle-like unit. He received the Nobel Prize in Physics in 1918 for his quantum theory after it had been successfully applied to the photoelectric effect by Einstein. Planck died in Göttingen, Germany in 1947.

Planck showed the following equation concerning the energy of a light to justify his quantum hypothesis.

$$E = h\nu$$

Here, $h$ denotes Planck's constant and $\nu$ the frequency of the radiation, respectively. The value of $h$ is as follows:

$$h = 2\pi\hbar = 6.626 \times 10^{-34} \, \text{J} \cdot \text{S}$$

In 1905, Einstein published the *light quantum hypothesis*, which supports that a light is both wave and particle. The particle of a light is called the *photon*. It is well known that the reason Einstein received the Nobel prize is not the theory of relativity but the light quantum hypothesis.

Planck's and Einstein's hypotheses produced the dispute of a light's *wave-particle duality*, because the duality cannot be interpreted by classical physics. Consequently, Planck's and Einstein's hypotheses are paradoxical, and the fact led to the birth of quantum mechanics.

Later, the research of physics moved to the explanation of the internal structure of an atom. The question is: What are atoms made of?

**Rutherford**

In 1911, Rutherford discovered the *atomic nucleus*. He performed the famous gold foil experiment which hit the gold foil with fast $\alpha$-particles. The experiment involved the firing of radioactive particles through minutely thin metal foils (notably gold) and detecting them using screens coated with zinc sulfide (a scintillator).

Rutherford found that although the vast majority of particles passed straight through the foil approximately 1 in 8000 were deflected, meaning that most of the atom was made up of empty space.

The result of the experiment led him to consider that atom consists of positive charged part and negative charged electrons around the positive charged part, and he named the positive charged part a nucleus.

**Ernest Rutherford (New Zealand: 1871-1937)**

Ernest Rutherford was born in Nelson, New Zealand on 1871. After studying at New Zealand, he moved to the University of Cambridge for his postgraduate study in 1895-1898. In 1898, he became the professor of McGill University in Canada. In 1911, he discovered the atomic nucleus and developed a model of the atom that was similar to the solar system. He received the Nobel Prize in Chemistry in 1908. Rutherford died in Cambridge, England on 1937.

In 1919, Rutherford also discovered a *proton* which is the element of a nucleus. He noticed that when alpha particles were shot into nitrogen gas, his scintillation detectors showed the signatures of hydrogen nuclei. The only place this hydrogen could have come from was the nitrogen, and therefore nitrogen must contain hydrogen nuclei.

He thus suggested that the hydrogen nucleus, which was known to have an atomic number of 1, was an elementary particle. He named it the proton. From his results, it was shown that an atom has a structure.

**De Broglie**
In 1924, De Broglie published a theory of *matter wave*, which showed that matters can be interpreted as waves. He extended Planck's and Einstein's theories by introducing the notion of a matter wave. He claimed that just as electromagnetic radiation can have both wave and particle properties, the components of matter may have wave-like as well as particle-like characters.

Namely, all matters have a specific wave associated with them. De Broglie established a connection between matter waves and the motion of intra-atomic particles like electrons. He then related the waves associated with all matters and applied them to photons. He summarized all of his findings and was able to use his new theories to prove Planck's law for blackbody radiation.

## Pauli

In the same year, Pauli discovered that an electron has an inherent value called the *spin*, which is intrinsic angular momentum associated with elementary particles. It is a purely quantum mechanical concept without any analogue in classical physics. And it is one of the bases of quantum computing.

In 1925, Pauli proposed *Pauli's exclusion principle*, which says that no two electrons may occupy the same quantum state simultaneously. It means that two identical electrons cannot be in the same energy level in the same atom.

## Heisenberg

In 1925, Heisenberg proposed the *matrix mechanics* as a foundation for quantum mechanics; see Heisenberg [47]. He described the *Heisenberg equation*, formalizing the relationship between the *observable*, which can be seen as a measurable physical value and the energy. As his theory is based on a matrix, it is now called the matrix mechanics. Note that Born, who was a teacher of Heisenberg, showed that Heisenberg's theory can be simplified by means of a matrix and that Heisenberg's theory became the present form, i.e., matrix mechanics.

Matrix mechanics assumes that all observables are expressed by matrices whose elements have indexes with two different energy levels. The set of eigenvalues of the matrix can be interpreted as the set of all possible values that observables can have. The eigenvalues are real since Heisenberg's matrices are hermitian. If an observable is measured and the result is a certain eigenvalue, then the corresponding eigenvector is the state of the quantum system immediately after the measurement.

**Max Born (Germany: 1882-1970)**

Max Born was born in Breslau, Poland on 1882. He entered the University of Göttingen in 1904. He, jointed with Heisenberg and Jordan, formulated quantum mechanics in the matrix version in 1927. He was awarded the Nobel Prize in Physics in 1954 for his fundamental research in quantum mechanics. Born died in Göttingen, Germany on 1970.

### Schrödinger

In 1926, Schrödinger developed *wave mechanics*, which is an alternative foundation for quantum mechanics. He described the properties of an electron by using the *Schrödinger equation*. Later, he proved that wave mechanics and matrix mechanics are mathematically equivalent.

In 1927, Heisenberg published the *uncertainty principle*, which can be seen as one of the fundamental concepts in quantum physics; see Heisenberg [48]. It states that we cannot simultaneously assign values of the pairs of the position and momentum, with full precision.

Consequently, attempting to measure a particle's position to the highest degree of accuracy, leads to an increasing uncertainty in the measurement of particle's momentum to an equally high degree of accuracy.

The uncertainty principle can be interpreted in matrix mechanics as follows. If two observables are measured simultaneously, then the state of the system should collapse to a common eigenvector of the two observables. However, eigenvectors in most matrices are not generally equal. Thus, most observables cannot be measured precisely at the same time.

In the same year, Pauli succeeded in formalizing the notion of *spin* in an electron by the Schrödinger equation and Heisenberg equation; see Pauli [65]. The spin can be interpreted as a fixed build-in angular momentum. Note also that the concept of spin plays an important role in quantum computers.

### Dirac

In 1928, Dirac proposed the *relativistic quantum mechanics* by unifying quantum mechanics and special theory of relativity. Dirac described the properties of an electron by the *Dirac equation*.

Dirac wanted an equation for the electron that would be consistent with special theory of relativity and quantum mechanics, and modeled the frequency spectrum of hydrogen. The required equation which Dirac discovered is the Dirac equitation. Dirac also predicted the existence of antiparticles like the positron based on his equation. In 1932, Anderson discovered the positorn from cosmic rays in a cloud chamber.

### Yukawa

In 1935, Yukawa, who is the first Japanese person to win the Nobel Prize, proposed a theory of *meson*. He predicted the existence of an elementary particle whose mass is between that of an electron and proton. The elementary particle is now called *pion*.

**Hideki Yukawa (Japan: 1907-1981)**

Hideki Yukawa was born in Tokyo, Japan on 1907. After studying physics at Kyoto Imperial University, he became Lecturer there in 1929. He received Ph.D. degree in 1938 from Osaka Imperial University, Japan. He published a theory of meson in 1935. Later, the elementary particle whose existence was predicted by Yukawa was discovered by Powell in 1947. He received the Nobel Prize in Physics in 1949. Yukawa died in Kyoto, Japan on 1981.

### Anderson and Neddermeyer

In 1936, Anderson and Neddermeyer discovered a meson, but it was different from the one Yukawa predicted its existence. The meson they discovered is now called the *muon*.

**Left: Seth Neddermeyer (USA: 1907-1988)**
**Right: Carl David Anderson (USA: 1905-1991)**

Seth Neddermeyer was born in Richmond, Michigan, USA on 1907. He received B.A. from Stanford University in 1929. He received Ph.D. degree at Caltech in 1935. He discovered the muon, with Anderson in 1936. He joined the Manhattan Project which developed atomic bombs in 1943-1946. Neddermeyer died in Seattle, Washington, USA on 1988.

Carl David Anderson was born in New York, USA on 1905. He received B.S. in 1927 and Ph.D. in 1930 from Caltech. He discovered the positron in 1932, and received the Nobel Prize in Physics in 1936 for the discovery, with Hess. He also discovered the muon, with his student Neddermeyer in 1936. Anderson died in San Marino, California, USA on 1991.

## Powell

In 1947, Powell discovered the pion which is the meson whose existence was pointed out by Yukawa in 1935. In 1950, Powell was awarded the Nobel Prize in Physics. In the previous year, Yukawa had received the award for his work on the meson theory.

**Cecil Frank Powell (England: 1903-1969)**

Cecil Frank Powell was born in Tonbridge, England on 1903. He received Ph.D. in Physics in 1927 from Cambridge University. He received the Nobel Prize in Physics in 1950 for his development of the photographic method of nuclear processes and his discoveries regarding mesons made with this method. Powell died in Valsassina, Italy on 1969.

## Hahn

In 1938, Hahn discovered the *nuclear fission*, which led to the ending of the World War II. From the fact that a slow neutron can be captured by uranium-235 (U-235) nucleus and radium whose mass is smaller than that of uranium is produced, he thought that the nucleus of an atom splits into parts.

Further experiments revealed that the U-235 fission yielded a very amount of energy, and that it yielded at least two neutrons per neutron absorbed in the interaction. Hahn realized that this made possible a chain reaction with an unprecedented energy yield. Then, nuclear fission can be seen as one of the bases of atomic energy.

**Otto Hahn (Germany: 1879-1968)**

Otto Hahn was born in Frankfurt am Main, Germany on 1879. From 1897, he studied
chemistry at the University of Marburg, Germany, and received his doctorate in 1901.
He discovered the nuclear fission, with Meitner and Fritz Strassmann in 1938, and
won the Nobel Prize in Chemistry in 1944. During the World War II, he served in the
army for chemical warfare. Due to his work, he is often called 'the father of nuclear
chemistry'. Hahn died in Göttingen, West Germany on 1968.

Later, the nuclear fission of uranium was applied to the atomic bombs in USA,
and led to the atomic bombings to Hiroshima and Nagasaki in 1945. Note that,
in connection with their development in which vast amounts of computation were
needed, von Neumann proposed a theory of computer. However, the implementations of computers based on his idea have been done after the World War II.

In this way, theoretical foundations for quantum mechanics have been established
by the 1930's. In addition, the structure of an atom was clarified theoretically as well
as experimentally by the mid of the 1940's.

That is, atoms are basically composed of three elementary particles, i.e., protons, neutrons and electron. The nucleus consists of the protons and neutrons. Protons have a positive charge, neutrons have no charge and electrons have a negative
charge, respectively.

Electrons reside in orbitals around the nucleus. The electron was discovered by
Thomson in 1897 and the proton was discovered by Rutherford in 1919, as noted
above.

**Joseph John Thomson (U.K.: 1856-1940)**

Joseph John Thomson was born in Manchester, U.K. on 1856. He received B.A. in mathematics in 1880 and M.A. in 1883 from Trinity College, Cambridge. He was Cavendish Professor of Experimental Physics at Cambridge in 1884. He discovered the electron in a series of experiments to investigate the nature of electric discharge in a high-vacuum cathode-ray tube in 1897, and was awarded the Nobel Prize in Physics in 1906 for the discovery of the electron and or his work on the conduction of electricity in gasses. He also found the stable isotopes by the experiments with cathode rays in 1911. Thomson died in Cambridge, U.K. on 1940.

## Chadwick

The neutron was discovered by Chadwick in 1932. The existence of neutron was predicted by his teacher Rutherford in 1920.

**James Chadwick (U.K.: 1891-1974)**

James Chadwick was born in Bollington, U.K. on 1891. He entered in Manchester University on 1908 and received M.Sc. degree in 1913. He became professor of Physics of the University of Liverpool, U.K. in 1935. He discovered the neutron in 1932, and this was useful for the understanding of the nuclear fission of uranium 235. He received the Nobel Prize in Physics in 1935 for his discovery of the neutron. During the World War II, he joined the Manhattan Project. Chadwick died in Cambridge, U.K. on 1974.

In 1932, Chadwick studied the beryllium radiation with an ionization and a cloud chamber. By kinematics, he was able to determine the velocity of the protons. Through conservation of momentum techniques, he could determine that the mass of the neutral radiation was almost exactly the same as that of a proton.

Then, he had conclusive proof of the existence of the neutron. Without Chadwick's discovery of neutron, Hahn would not explain nuclear fission. Also note that his discovery was quickly accepted. For example, Heisenberg showed that the neutron could not be a proton-electron pair, but had to be its unique particle. This implied that the third piece of the atom was found.

In the 1930's, the research of mathematical foundations for quantum mechanics has also been conducted. In 1930, Dirac presented an alternative formulation of quantum mechanics by means of the *bra-ket notation*; see Dirac [29].

The bra-ket notation is concise and convenient way to describe quantum states. It became the standard notation in quantum mechanics. We will explain the details of the bra-ket notation in Chapter 3, since it is often used in the literature on quantum computing.

### von Neumann

In 1932, von Neumann advanced a mathematical foundation for quantum mechanics based on the Hilbert space in von Neumann [82]. His theory of quantum mechanics is useful to the understanding of quantum computers.

He also proposed *quantum logic*, which is a logic for quantum mechanics, in 1936; see Birkhoff and von Neumann [14]. Quantum logic is regarded as a logic which deals with the propositions determined by measurements. It can be used to give a logical (also philosophical) foundation for quantum mechanics. We will discuss quantum logic in connection with quantum computing in Chapter 6.

Obviously, von Neumann contributed to both quantum mechanics and computers. He provided a first theoretical foundation for quantum mechanics using Hilbert space. He also played an important role for the design of current computers. However, it seems that he did not work out the possibility of realizing a quantum computer.

After the World War II, quantum mechanics has been further developed. However, the recent trends of quantum mechanics are not always directly connected with current quantum computers. Thus, we omit the survey of recent developments.

# Chapter 2
# Models of a Computer

**Abstract.** Chapter 2 surveys current computer models including Neumann-type computer, Turing Machine and Boolean algebra. Before looking at quantum computers, we have to understand these models properly.

## 2.1 Neumann-Type Computer

The first electronic computer called *ENIAC* (Electronic Numerical Integrator And Computer) was developed at University of Pennsylvania, USA in 1946 by Eckert and Mauchly; see Grier [43].

**Left: J. William Mauchly (USA: 1907-1980)**
**Right: J. Presper Eckert (USA: 1919-1995)**

John William Mauchly was born in Cincinnati, USA on 1907. He entered John Hopkins University in 1925. He received Ph.D. degree from the university in 1932. He became a professor of Physics at Ursinus College, USA, in1933. He is a physicist, and he designed the first computer, i.e., ENIAC in 1942, with Eckert. Mauchly died in Ambler, USA on 1980.

S. Akama, *Elements of Quantum Computing*,                                                    17
DOI: 10.1007/978-3-319-08284-4_2, © Springer International Publishing Switzerland 2015

John Persper Eckert was born in Philadelphia, USA on 1919. He studied at the University of Pennsylvania's Moore School of Electrical Engineering. He was an electrical engineer. He died in Bryn Mawr on 1995.

They also played an important role in the development of several computers. After the development of ENIAC, they designed EDVAC in 1951 and UNIVAC I in 1951. The latter was the first commercial computer.

### 2.1.1   ENIAC and EDVAC

We briefly describe the history of ENIAC. In 1943, the military commission on the new computer began. Mauchly was the chief consultant and Eckert was the chief engineer. Eckert was a graduate student studying at the Moore School when he met John Mauchly in 1943. It took the team about one year to design the ENIAC and 18 months and 500,000 tax dollars to build it. By that time, the war was finished.

ENIAC was completed by February, 1946. Its first task was to do computations for the development of a hydrogen bomb. ENIAC required 18,800 vacuum tubes, and used plug boards for communicating to the machine. It covered 1800 square feet of floor space, weighed 30 ton, and consumed 160 kilowatts of electrical power. There was a rumor that ENIAC caused brownouts in the city of Philadelphia.

ENIAC was designed for the specific purpose of computing. Therefore, ENIAC, which is different from modern computers for the general purpose, is called *wired program computer*. According to Grier [43], ENIAC is better described as a collection of electronic adding machines and other arithmetic units, which were originally controlled by a web of large electrical cables.

In 1948, von Neumann made several modifications to the ENIAC. The reason lies in the fact that it had performed arithmetic and transfer operations concurrently, making programming difficult. Von Neumann suggested that switches control code selection so pluggable cable connections could remain fixed. He also added a converter code to enable serial operation.

The architecture of modern computers is based on stored program type, which was proposed by von Neumann in 1945; see von Neumann [83]. Now, the computer based on the architecture is therefore called the *Neumann-type computer*. Most computers we are now using are Neumann-type.

Von Neumann proposed a model for the Neumann-type computer for the new computer *EDVAC* (Electronic Discrete Variable Automatic Computer), but the development was delayed due to several problems including the patent. The first Neumann-type computer was completed at Cambridge University in 1949, and EDVAC was completed at University of Pennsylvania in 1951.

EDVAC was also designed by Eckert and Mauchly. They started the work two years before the operation of ENIAC. In EDVAC, the program is stored inside it. EDVAC used binary rather than decimal numbers, and the arithmetic units were simplified. In this sense, we should say that EDVAC is a *first* computer.

Von Neumann called a computer an automatic computing system and defined it as follows:

An automatic computing system is a (usually highly composite) device, which can carry out instructions to perform calculations of a considerable order of complexity—e.g. to solve a non-linear partial differential equation in 2 or 3 independent variables numerically.

Modern computers in fact realize the definition. But, by using modern computers we can also perform other computations such as image processing, speech processing, knowledge processing as well as numerical computation. The Neumann-type computer has the following features.

---

(1) Programs and data are stored in a computer.

(2) Computation is performed sequentially by the instructions of a program.

---

Here, the *program* is a set of machine instructions indicating the procedure of a computation. Data are objects, e.g., numbers, characters and strings, needed for computation.

The feature (1) is the reason the Neumann-type is also called stored program type. The feature (2) denotes that a computer requires program and data for its computations.

## 2.1.2 Architecture of Neumann-Type Computers

Computers which we are usually using in the present day are classified as the Neumann-type. Computers which are not he Neumann type are called the non-Neumann-type. Thus, quantum computer is understood as one of the non-Neumann-type computers.

As you know, a computer consists of the *hardware* and the *software*. A hardware is itself the device of a computer, and a software is a set of programs which makes it work. In this regard, a computer cannot work without software.

Hardware consists of the following units.

---

- control unit
- arithmetic unit
- main storage unit
- secondary storage unit
- input unit
- output unit

---

The structure of hardware can be described as Figure 2.1.

*Input unit* takes the input and converts it into the binary form. It is formed by input devices like keyboard and mouse attached to the computer.

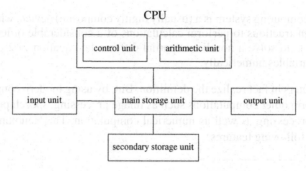

**Fig. 2.1** Hardware

*Output unit* gives the result of computation. The Visual Display Unit (VDU) and the printer are common output unit.

Input unit and output unit are gathered, and is called the *input-output unit*. Together with main storage unit and secondary unit is called the *storage unit*.

*Control Processing Unit* (CPU) generally consists of the control unit and arithmetic unit. The control unit controls a computer including other units. The arithmetic unit computes by using data stored in the storage unit.

*Main storage unit*, also called the *memory*, is the unit which stores both programs and data.

*Secondary storage unit* stores both programs and data, which need not perform immediately.

In the Neumann-type computers, programs and data for particular computation are stored in the main storage unit. Since computers can understand only the *machine language* whose instructions are described as a sequence of *binary digits*, we have to write a program in some *programming language* like the C-language and transform it into the one in the machine language to execute the program in a computer.

Programs and data are taken from the main storage unit when executing. For this purpose, however, it is necessary to recognize where they are located in the main storage unit. The location where programs or data are stored is specified by the *address*. Thus, the address enables us to access the main storage unit. The CPU performs actual computation. In other words, CPU decodes and executes instructions sequentially.

The above is the outline of the Neumann-type computer. We can perform complicated computation by the combination of simpler operations in binary digits in the Neumann-type computer.

## 2.2 Turing Machine

Many people seem to think that when a computer was invented its computational model was also invented. But it is not true. In fact, there have been many computational models before a computer was developed. It is observed that such computational models have been established by the 1930's.

Historically, the implementation of a computer has been done after the World War II. Notice that, however, several projects for computers have been progressed in the period. And important figures like von Neumann and Turing have joined such projects. The fact seemed to be very important.

**Alan Turing (U.K.: 1912-1954)**

Alan Turing was born in London, U.K. on 1912. He studied Mathematics at King's College, Cambridge. In 1935, he was elected as a Fellow of the college. He proposed one of the most important theories of computation called the Turing Machine in 1936-1937. During World War II, he worked on decoding German's code. In 1946, he wrote a paper on the design of ACE (Automatic Computing Engine), which is a stored program computer. In 1950, the Pilot Ace, which is a small version of ACE, ran for the first time. He became a Reader of Manchester University in 1948. In 1950, he proposed a test, now called the Turing test, which is a test of a computer's ability to think. Turing died in Wilmslow, U.K. on 1954.

As mentioned above, EDVAC is regarded as the first stored program computer, which was designed by von Neumann. For this reason, the modern computer is called the Neumann-type. However, Turing provided a *complete* description of the stored program computer for ACE. Therefore, there may be a sense in which Turing is one of the founders of a computer.

Now, we review researches of computation in theoretical settings in some detail. In the 1930's, the problem of formalizing *computability* became important. Namely, this is the problem of clarifying the concept of a function which can be *concretely* computed.

A theory which formalizes the notion of computability is generally called the *theory of computation*; see Boolos and Jeffrey [16]. It studies the nature of computation. By the 1930's, the major theories of computation have emerged in the literature. They include Turing's *Turing machine* (TM), Church's $\lambda$-*calculus* [20], and Kleene's *recursive function* [56]. And it is well known that they were proved to be equivalent; see Kleene [56] for details.

Turing Machine is one of the popular theories of computation, and it was also extended for the first model for quantum computers. Other theories of computation are not always popular for us. It is thus necessary to give an exposition of Turing Machine here. Turing Machine was proposed by Turing in 1936; see Turing [79]. Turing is also famous that he decoded German's code during the World War II. Turing Machine was used to prove the undecidability of first-order logic.

Turing Machine was originally developed not for a computer but for computation in a desk. Namely, this is a "machine" that only needs a pencil and a paper. In this sense, Turing Machine is an abstract machine rather than a concrete machine like modern computer. Although Turing Machine has a computational mechanism different from the one of modern computers, i.e., Neumann-type computer, it gave a great influence in the birth of the computer. It is no doubt that von Neumann's model for a computer was influenced by Turing Machine.

## 2.2.1   Structure of Turing Machine

We turn to the structure of Turing Machine. It has the structure depicted in Figure 2.2. Namely, it consists of three components, i.e., tape, head and control unit.

**Fig. 2.2** Turing Machine

Here, the *tape* has infinite length in both left and right direction and distinguished by cells. We can write (only) a symbol including the blank in each square.

The *head* scans the square which can be read and written. Therefore, the tape and head correspond to the input-output unit. The head can move left and right. The tape has an infinite states, and has a mechanism of the storage unit.

The *control unit* stores a current state of Turing Machine. It has a finite memory and does the operation which corresponds to the symbol written in the cell reading from the tape.

Turing Machine can perform computation autonomously. In other words, it can do the computation by changing the state of the control unit and the symbols in the tape. In this regard, Turing Machine can be viewed as a computer. In fact, the control unit of Turing Machine corresponds to CPU in a modern computer.

Here are the procedures for the computation in Turing Machine.

---

(P1) Rewrite the symbol scanned by the head
(P2) Move the head left one cell
(P3) Move the head right one cell
(P4) Finish a computation

---

In (P1)-(P3), the control unit can change the internal state. In this sense, these operations are crucial to the computation in Turing Machine.

### 2.2.2 Formal Definition of Turing Machine

There are several definitions of Turing Machine in the literature. We give a standard one here. A Turing Machine can be defined by $M = \langle \Sigma, Q, \delta \rangle$, where:

---

- $Q$ is a set of internal states. The initial state $q_0 \in Q$ and the final state $q_f \in Q$, and $q_0 \neq q_f$.
- $\Sigma$ is the finite alphabet, i.e., the finite set of tape symbols (Blank symbol $B \in \Sigma$).
- $\delta : \Sigma \times Q \to \Sigma \times Q \times \{L, R\}$ is the state transition function, where $L$ denotes the left and $R$ the right, respectively.

---

Here, $\delta$ is a function describing the transition of the state before a computation and the state after the computation.

A computation in Turing Machine $M$ is formally described as a *configuration*. A configuration completely represents tape symbols, the position of the head, and states in the Turing Machine $M$. And it is supposed that in a cell a finite symbol other than the blank is always written.

The initial configuration is the configuration in which the head is at the cell 0 and the state is the initial state $q_0$. The initial configuration an input $x \in (\Sigma - B)^*$, when $x$ is written in the position of the tape $0, 1, 2, \dots$ and other cells are empty. Here, $A^*$ denotes the set of any sequence of elements.

Turing Machine $M$ stops in the input $x$, when its state becomes the final state $q_f$. If Turing Machine $M$ stops, then its output the left to right string of symbols which is not a blank. If Turing Machine $M$ stoops for any input, the function $(\Sigma - B)^* \to \Sigma^*$ is computed.

For example, the Turing Machine which computes the function $succ(x) = x + 1$ is defined as follows:

$$M = \langle \{1\}, \{q_0, q_1, q_2\}, \{q_0 1 L q_1, q_1 B 1 q_2\} \rangle$$

The configuration of the function $succ(1)$ is as follows:

(1) $q_0 1$

(2) $q_1 B 1$

(3) $q_2 1 1$

(1) is the configuration in which the computation is started. Firstly, the argument 1 of $succ(1)$ is written to the cell which the head scans. The initial state is $q_0$.

From (1), the state transition function $q_0 1 L q_1$ can be applied. Namely, we can change the internal state as $q_1$ and move the head left one. As a result, the configuration (2), i.e., $q_1 B 1$ can be obtained. Here, the position of the head is $B$ (blank).

From (2), the state transition function $q_1 B 1 q_2$ can be applied. Namely, we can change the internal state as and rewrite $B$ as 1. As a result, the configuration (3), i.e., $q_2 11$ can be obtained. Since $q_2$ is the final state, the computation finishes.

The contents of the tape when the computation finishes, i.e., 11 is the result of computation. It follows that we compute $0 + 1$ in decimal system because Turing Machine is based on the unary numeral system, and the result is $11_1 = 1_{10}$.

The Turing Machine described above computes a particular function. As its computation is performed deterministically, it is called the *deterministic Turing Machine*. The Turing Machine which can simulates all Turing Machines called the *universal Turing Machine*.

If there exists a Turing Machine which can compute partial function $f$, we call $f$ *Turing computable*. Church considered a *computable function* as the function which is Turing computable, and this idea called the *Church's thesis* (CT) or Church-Turing thesis.

It says that a computable function is the function which can be computed by a Turing Machine, that is, the Turing computable function. But Church's thesis cannot be proved since it is not a theorem but an empirical conjecture.

Observe also that the computable functions can be defined by $\lambda$-calculus or recursive function. Therefore, the functions which can be computed by these theories are also regarded as computable functions.

There are several variants of Turing Machines. Here, we mention some of them briefly. A *non-deterministic Turing Machine* is the Turing Machine whose computation is performed non-deterministically. Observe that the value of the state transition function of non-deterministic Turing Machine is not unique, and its configuration can be described as a tree.

A *probabilistic Turing Machine* is the Turing Machine) whose state transition function obeys a probabilistic distribution and the computation is probabilistically described. Thus, probabilistic Turing Machine can be classified as non-deterministic Turing Machine. There is a possibility of having different results in a probabilistic Turing Machine, since its internal states are probabilistically transited.

A quantum Turing Machine is, as will be described later, can be interpreted as a special form of probabilistic Turing Machine.

## 2.3 Boolean Algebra

For constructing the hardware of a computer, it is very useful to use the *Boolean algebra*, which is also called the logical algebra or Boolean lattice. Thus, Boolean algebra is more important than Turing Machine in the practice of the hardware of a computer; see Mendelson [62].

An *algebra* can be defined as a structure and the operations with respect to it. There are in fact many algebras studied in the literature. For instance, group, ring and field are algebras. And these algebras are the subjects of research in mathematics.

Boolean algebra was proposed by George Boole in the 19th century, independently to propositional logic. Boole proposed it to formalize the method for computing human's thoughts; see Boole [15].

There are several ways to formalize Boolean algebra. For example, Boolean algebra can be formalized lattice-theoretically; see Birkhoff [13] and Davey and Priesley [23]. However, we here show a standard formulation generally used in computer science. The symbols in Boolean algebra are as follows:

---

$\bar{A}$ (NOT)

$A \cdot B$ (AND)

$A + B$ (OR)

$A = B$ (EQUAL)

---

Here, $A$ and $B$ denote a formula called a *proposition*, whose truth is determined, i.e. true (1) or false (0). 0 and 1 are special propositions which express truth and falsity, respectively.

$\bar{A}$ denotes the *negation*, $A \cdot B$ denotes the *conjunction*, $A + B$ denotes the *disjunction*, $A = B$ denotes *equivalence*, respectively.

Note that "·" may be omitted. In the literature, we may find different notations. For instance, $\bar{A}$ is written as $-A$ and $A \cdot B$ as $A \times B$.

In Boolean algebra, 1 and 0 are also used as *truth-values*, representing truth and falsity. Formulas in Boolean algebra are described as the form of equations. For example, the formula of the form $A = 1$ expresses that the proposition $A$ is true and the formula of the form $A = 0$ expresses that the proposition $A$ is false. We write $A = B$ to mean that the truth-values of $A$ and $B$ are equal and $A \neq B$ to mean that the truth-values of $A$ and $B$ are not equal.

## 2.3.1 Axiomatization of Boolean Algebras

One way to formalize Boolean algebra is the *axiomatic system*, which may not be standard for computer science. An axiomatic system can be defined by means of a set of *axioms* and *rules of inference*.

An axiom is a formula which is assumed to be true. A rule of inference stipulates the method how inferences should be done. A formula which is obtainable from the axioms using the rules of inference is called the *theorem*. In fact, axioms are also theorems.

The following is one of the axiomatic systems for Boolean algebra.

### Axiomatic System for Boolean Algebra

(B1) $A + B = B + A$

(B2) $A \cdot B = B \cdot A$

(B3) $A + (B \cdot C) = (A + B) \cdot (A + C)$

(B4) $A \cdot (B + C) = (A \cdot B) + (A \cdot C)$

(B5) $A + 0 = A$
(B6) $A \cdot 1 = A$
(B7) $A + \bar{A} = 1$
(B8) $A \cdot \bar{A} = 0$
(B9) $0 \neq 1$

A rule of inference in Boolean algebra is only the *rule of substitution*.

Each axiom of Boolean algebra describes the property which should be satisfied. (B1) and (B2) are the *commutative law* for disjunction and conjunction, respectively.

(B3) and (B4) are the *distributive law* about conjunction and disjunction.

(B5) and (B6) are the *absorption law* for disjunction and conjunction, respectively.

(B7) is called the *law of excluded middle*, which states that the truth-value of a proposition is either 1 or 0.

(B8) is called the *law of non-contradiction*, which states that the truth-value of a proposition is not both 1 and 0. Since (B8) is obvious, it may be not included as an axiom.

(B9) shows that truth and falsity are different propositions.

The following formulas are representative theorems of Boolean algebra.

(1) $\bar{\bar{A}} = A$
(2) $\overline{A + B} = \bar{A} \cdot \bar{B}$
(3) $\overline{A \cdot B} = \bar{A} + \bar{B}$
(4) $A + (B + C) = (A + B) + C$
(5) $A \cdot (B \cdot C) = (A \cdot B) \cdot C$

(1) is called the *double negation law*. (2) and (3) are called *de Morgan's law*. (4) and (5) are called the *associativity law*.

In Boolean algebra, not all logical symbols are needed. From theorems (1), (2) and (3), we can perform a computation in Boolean algebra only by negation and conjunction or by negation and disjunction.

As is well-known, there is an intimate connection of Boolean algebra, *propositional logic* and *set*. Logical symbols in propositional logic have the counterpart in Boolean algebra. However, implication $\rightarrow$, which is a basic logical symbol in propositional logic, is not generally used in Boolean algebra[1].

It is also possible to interpret the algebra based on *power set*, which is a set of all subsets in a set. It can be viewed as Boolean algebra. In fact, union $\cup$ corresponds to $+$, intersection $\cap$ to $\cdot$, complement $^c$ to $^-$, respectively. $\emptyset$ is 0 and $E$ is 1, where *emptyset* denotes the empty set and $E$ denotes the universal set, respectively.

In Boolean algebra, as in propositional logic, some other logical symbols are also used. $A \mid B$ is called *Sheffer's stroke*, also known as *NAND*, which is defined as follows:

---

[1] Implication $A \rightarrow B$ is defined by $\neg A \vee B$, where $\neg$ denotes negation and $\vee$ disjunction, respectively.

$A \mid B = \overline{A \cdot B}$

$A \downarrow B$ is called *Peirce's down arrow*, also known as *NOR*, which is defined as follows:

$A \downarrow B = \overline{A + B}$

$A \oplus B$ is called the *exclusive or* (XOR), whose definition is as follows:

$A \oplus B = A \cdot \bar{B} + \bar{A} \cdot B$

As will be clear, the above three logical symbols play an important role in designing the hardware of a computer.

### 2.3.2   Truth-Value Table

Logical symbols in Boolean algebra can be interpreted by the so-called *truth-value table*. In other words, we can understand the meanings of logical symbols by means of the truth-value table which gives the truth-value of the whole formula from the truth-values of its subformulas.

We here describe how to give the truth-value table for a formula.

First, from the left most row, simpler subformulas are written in turn, and in the right most row the formula evaluated is written.

Next, assign possible truth-values to subformulas as columns for computing the truth-value of the formula evaluated.

In computing the truth-value of a formula, the truth-value tables for basic logical symbols are utilized.

The truth-value table for 'NOT' is presented as Table 2.1.

**Table 2.1** Truth-Value Table of NOT

Consequently, the interpretation of 'NOT' reads[2]:

$\overline{A} = 1$ iff $A = 0$.
$\overline{A} = 0$ iff $A = 1$.

Thus, $\overline{A}$ is true iff $A$ is false. Actually, we can give the interpretation by one of them due to the law of excluded middle.

The truth-value table of 'AND' and 'OR' is given as Table 2.2.

Thus, the interpretation of 'AND' reads:

---

[2] Here, 'iff' is shorthand for 'if and only if'.

**Table 2.2** Truth-Value Table of AND and OR

| A | B | $A \cdot B$ | $A + B$ |
|---|---|---|---|
| 1 | 1 | 1 | 1 |
| 1 | 0 | 0 | 1 |
| 0 | 1 | 0 | 1 |
| 0 | 0 | 0 | 0 |

$A \cdot B = 1$ iff $A = B = 1$.
$A \cdot B = 0$ iff $A = 0$ or $B = 0$.

In other words, $A \cdot B$ is true iff both $A$ and $B$ are true.
The interpretation of 'OR' is as follows:

$A + B = 1$ iff $A = 1$ or $B = 1$.
$A + B = 0$ iff $A = B = 0$.

We can intuitively interpret that $A + B$ is true iff either $A$ or $B$ is true
The truth-value tables of 'NAND', 'NOR' and 'XOR' is given as Table 2.3.

**Table 2.3** Truth-Value Table of NAND, NOR and XOR

| A | B | $A \mid B$ | $A \downarrow B$ | $A \oplus B$ |
|---|---|---|---|---|
| 1 | 1 | 0 | 0 | 0 |
| 1 | 0 | 1 | 0 | 1 |
| 0 | 1 | 1 | 0 | 1 |
| 0 | 0 | 1 | 1 | 0 |

'NAND' is interpreted in the following way:

$A \mid B = 0$ iff $A = B = 1$.
$A \mid B = 1$ iff it is not the case that $A = B = 1$.

Intuitively, $A \mid B$ is false iff both $A$ and $B$ are true.
The interpretation of 'NOR' is as follows:

$A \downarrow B = 1$ iff $A = B = 0$.
$A \downarrow B = 0$ iff it is not the case that $A = B = 0$.

The intuitive interpretation shows that $A \downarrow B$ is true iff both $A$ and $B$ are false.
We can give the interpretation of 'XOR' as follows:

$A \oplus B = 1$ iff $A \neq B$.
$A \oplus B = 0$ iff $A = B$.

The interpretation has the intuition that $A \oplus B$ is false iff the truth-values of $A$ and $B$ are equal.

Observe that other logical symbols can be defined by 'NAND' (or 'NOR'). This is the point in the hardware design in that these two logical symbols can simplify the representation of logical circuits. In fact, they are the universal gate.

We say that a formula is a *tautology* if it receives the truth-value 1 when any truth-values are assigned to its subformulas. Due to the completeness, the notions of tautology and theorem are equivalent in Boolean algebra.

We say that a formula is *satisfiable* if there is an assignment to the subformulas of a formula giving the truth-value 1 to the formula.

For instance, Table 2.4 is the truth-value table of the formula $A + \overline{A}$, called the law of excluded middle.

**Table 2.4** Truth-Value Table of $A + \overline{A}$

| $A$ | $\overline{A}$ | $A + \overline{A}$ |
|---|---|---|
| 1 | 0 | 1 |
| 0 | 1 | 1 |

Here, from the rows of $A$ and $\overline{A}$, we can compute the truth-value of $A + \overline{A}$. Looking at the row of $A + \overline{A}$, they are all 1. It can be shown that the formula $A + \overline{A}$ is a tautology.

Boolean algebra can also be formalized as the form of *Boolean function*. Boolean algebra can serve as the basis of the hardware of a computer, as discussed in the next subsection.

### 2.3.3 Boolean Algebra and Hardware

One of the important areas in which Boolean algebra is successfully applied is undoubtedly the hardware of a computer. The use of Boolean algebra led to the simple hardware design of a computer. It is thus necessary for those working on computer to computer science understand Boolean algebra.

In the design of the hardware, we employ a *logic circuit*. It is an electric circuit whose output depends on the input and constructed by the combination of *logic gates*. It is considered as a simple circuit with one or two inputs and one output, and does a basic logical operation.

It is useful to visualize logic gates in design. A notation for the logic gates representing the basic logic circuit are stipulated as the standards called *MIL* as Figure 2.3.

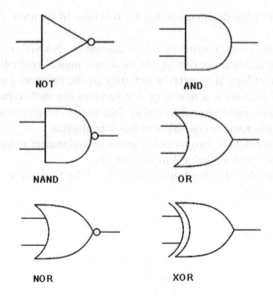

**Fig. 2.3** Logic Gates

Here, the lines from left to symbols denotes an input and the line from symbols to right an output, respectively. There are other notations for logic gates.

In designing a logic gate, it is important to reduce logical equations in Boolean algebra to a circuit with fewest number of gates.

For example, we consider the following logical equation.

$$X = A \cdot B \cdot C + A \cdot \overline{B} \cdot C + \overline{B} \cdot \overline{C} + \overline{A} \cdot \overline{B}$$

It can be reduced by means of Boolean algebra by means of several laws. The reduction can be done mechanically.

First, we reduce it by expanding all terms by multiplying by 1. The result is as follows:

$$X = A \cdot B \cdot C + A \cdot \overline{B} \cdot C + A \cdot \overline{B} \cdot \overline{C} + \overline{A} \cdot \overline{B} \cdot \overline{C} + \overline{A} \cdot \overline{B} \cdot C + \overline{A} \cdot \overline{B} \cdot \overline{C}$$

By taking out the common factor, we have:

$$X = A \cdot B \cdot C + \overline{B} \cdot (A \cdot C + A \cdot \overline{C} + \overline{A} \cdot \overline{C} + \overline{A} \cdot C + \overline{A} \cdot \overline{C})$$

Next, it can be reduced by grouping terms and taking out the common factor.

$$X = A \cdot B \cdot C + \overline{B} \cdot (A \cdot (C + \overline{C}) + \overline{A} \cdot (\overline{C} + C + \overline{C}))$$

Next, we can simplify by it as follows:

$$X = A \cdot B \cdot C + \overline{B} \cdot (A + \overline{A})$$

Finally, we can obtain the simplest form:

$$X = A \cdot B \cdot C + \overline{B}$$

The reduction of logical equations can be done by using the so-called *Karnaugh map*. We omit the details of it here, however.

# Chapter 3
# Quantum Mechanics

**Abstract.** Chapter 3 introduces quantum mechanics. After looking at the basic ideas of quantum mechanics, wave mechanics, matrix mechanics, and the uncertainty principle are explained. Since the materials in this chapter are mathematically advanced, the readers who are not interested in them can skip and proceed to Chapter 4.

## 3.1 Basics of Quantum Mechanics

Before studying quantum computers, we must learn its underlying basis, i.e., quantum mechanics. For comprehensive exposition, the reader is referred to von Neumann [82], Dirac [29] and Green [42].

### 3.1.1 History

As described in Chapter 1, in 1900, Planck proposed the quantum hypothesis which claims a light is a particle. Since a light was considered as a wave, the quantum hypothesis led to the wave-particle duality. This is a surprising fact in view of classical physics.

In 1924, De Broglie showed a theory of *matter wave*. According to his hypothesis, every particle, like photon, can be also interpreted as wave in that it has a specific wave associated with it. The hypothesis that all particles have both wave-like and particle-like properties is called the *wave-particle duality*.

**Louis de Broglie (France: 1892-1987)**

S. Akama, *Elements of Quantum Computing*,
DOI: 10.1007/978-3-319-08284-4_3, © Springer International Publishing Switzerland 2015

Louis de Broglie was born in Dieppe, France on 1892. After studying at the Lycee Janson of Sailly, he passed his school-leaving certificate in 1909. He applied himself first to literary studies and took his degree in history in 1910. Then, he studied for a science degree, and gained in 1913. He proposed several important theories related to quantum mechanics. The theory of wave-particle duality was published in 1924. He also proposed Neutrino theory of light in 1934. He received the Nobel Prize in Physics in 1929. De Broglie died in Louveciennes, France on 1987.

Let $E$ be the energy, $\omega$ be the frequency of vibration, $p$ be the movement vector, and $k$ be the wave vector. Then, de Broglie's hypothesis can be described as follows.

$$E = \hbar\omega$$
$$p = \hbar k$$

De Broglie did not justify his hypothesis by actual experiment. In 1927, de Broglie's hypothesis was justified by Davisson and Germer's experiment of *diffraction*. Diffraction is the slight bending of wave (light) as it passes around the edge of an object. They shot electrons onto a nickel crystal, and found that the diffraction of the electron similar to waves diffraction against crystal (X-rays). This implies that the electron can be interpreted as a wave.

**Left: Clinton Davisson (USA: 1881-1958)**
**Right: Lester Germer (USA: 1896-1971)**

Clinton Davisson was born in Bloomington, USA on 1881. He entered the University of Chicago. He received B.S. degree in 1908 from the University of Chicago and Ph.D. in 1911 from Princeton University. He received the Nobel Prize in Physics in 1937, with Thomson. Davisson died in Charlottesville, USA on 1958.

Lester Germer was born in Chicago, USA on 1896. He served in World War I as a flight pilot. He worked as an assistant of Davisson at Bell Laboratory. Germer died in Gardiner, USA on 1971.

In 1927, Davisson and Germer performed the experiment which justifies de Broglie's hypothesis that the electron should show the properties of a wave as well as those of a particle.

However, in 1933, Compton discovered the phenomena convincing that light is made up of particles, i.e., photons, which is known as the *Compton effect*. It is the result of a high-energy photon colliding with a target, releasing loosely bound electrons from the outer shell of the atom or molecule. The scattered radiation showed a wavelength shift. However, it cannot be explained by wave theory. Namely, a light is shown to be a particle. There is evidence supporting Einstein's photon theory.

The wave-particle duality challenged physicists, because it cannot receive classical physical interpretation. On the one hand, the particulate is the property in which particles can be counted and their movement has a locus. On the other hand, both interference and diffraction support the fluctuation.

### 3.1.2   From Classical Mechanics to Quantum Mechanics

The solution to interpret the wave-particle duality is to work out quantum mechanics which can model the behavior of matters in the micro world. In contrast, classical mechanics is a theory capable of formalizing the behavior of matters in the macro world.

As is well known, in classical mechanics, the movement of matters can be described by Newton's equation of movement, i.e.,

$$F = ma$$

where $a$ is the acceleration vector, $F$ is the force vector, and $m$ is the scalar, respectively.

By the equation of movement, the position and momentum can be determined simultaneously. This means that the position and velocity of a particle in past, present and future can be determined. Thus, *causality* holds in classical mechanics. But, in quantum mechanics, causality does not hold. Instead the Uncertainty Principle holds.

The purpose of quantum mechanics is to describe mathematically the movement of a particle in the micro world. In 1926, Schrödinger proposed the Schrödinger equation for an electron, and this is the origin of wave mechanics.

In 1925, Heisenberg formalized a similar theory in terms of matrix, known as matrix mechanics. Later, Schrödinger and Dirac proved that matrix mechanics and wave mechanics are mathematically equivalent.

## 3.2   Wave Mechanics

Generally, the exposition of quantum mechanics uses wave mechanics due to the fact that wave mechanics is easy to understand. However, we must also understand the basics of matrix mechanics, although the exposition based on it may be novel in general. The latter is of interest because theoretical foundations for quantum computers have the viewpoint similar to matrix mechanics.

In physics, various phenomena are described as a *physical system*. We can explain physical phenomena by showing the corresponding physical systems mathematically. In this sense, the behavior of a particle is formulated by the *particle system*. Here, the properties of a particle are given by factors like mass, momentum, coordinate and time. In general, a physical system can be expressed by using differential equations.

Classical mechanics can be simply formalized as the form known as the *Hamiltonian*. Such formalization is also possible by means of the *Lagrangian*. Now, we represent a particle system as the coordinate $q_1, q_2, \ldots$, momentum $p_1, p_2, \ldots$, and time $t$. If the particle's energy $H$ is expressed as the function of $q, p, t$, i.e.,

$$H = H(q_1, q_2, \cdots, p_1, p_2, \cdots, t)$$

Then the function $H(q_1, q_2, \cdots, p_1, p_2, \cdots, t)$ is called the *Hamilton function* of the particle system.

In the Hamiltonian, the velocity $\dot{q}_r$ of a particle $r$ can be obtained by differentiating the Hamilton function $H$ for $p_r$. Namely, the following holds.

$$\dot{q}_r = \frac{\partial H}{\partial p_r}$$

Here, the dot ˙ stands for the differential function. The equation of movement of the particle system is then described as follows:

$$\dot{p}_r = -\frac{\partial H}{\partial q_r}$$

In classical mechanics, the following holds.

$$\dot{p}_r = m_r \dot{q}_r$$

Observe that the law of conservation of mechanical energy, the sum $E$ of kinetic energy $T$ and potential energy $V$ is always constant. Here, $V$ is usually a function of the coordinate $V(q_1, q_1, \ldots)$. The kinetic energy $T$ is defined as follows:

$$T = \sum_r \frac{1}{2} m_r \dot{q}_r^2$$

We can write the equation of movement in the following way:

$$m_r = \frac{d\dot{q}_r}{dt} = -\frac{\partial V}{\partial q_r}$$

Here, $m_r$ represents the mass of an object at the $r$-th coordinate. Therefore, the Hamiltonian $H$ in classical mechanics can be expressed as follows:

$$H = \sum_r (\frac{1}{2} m_r \dot{q}_r^2) + V = \sum_r \frac{1}{2m_r} p_r^2 + V$$

### 3.2.1 Wave Function

In quantum mechanics, it is necessary to give an equation of movement to model the behavior of a particle which has the particle-wave duality. In 1926, Schrödinger proposed such an equation now called the *Schrödinger equation*, which is also called the wave equation.

**Erwin Schrödinger (Austria: 1887-1961)**

Erwin Schrödinger was born in Vienna on 1887. He studied at the University of Vienna from 1906 to 1910. In 1914, he achieved Habilitation. During the World War I he served as an artillery officer. He is a founder of wave mechanics and presented a basic equation, i.e., the Schrödinger equation proposed in 1926. He received the Noble Prize in Physics in 1933, with Dirac. He also showed a paradox known as Schrödinger's cat which is regarded as a paradox in quantum mechanics in 1935. Schrödinger died in Vienna on 1961.

A *wave function* is of the form $\Psi(q,t)$, which represents a possible state of a particle in time $t$ and position $q = (x, y, z)$. The Hamiltonian $H$ for the wave function can be defined as follows:

$$H = \frac{1}{2m}p^2 + V(q)$$

$$p = \frac{\hbar}{i}\nabla$$

$$\nabla = \left(\frac{\partial}{\partial x}, \frac{\partial}{\partial y}, \frac{\partial}{\partial z}\right)$$

Here, $V$ expresses the potential energy and $i$ the imaginary unit, respectively.

### 3.2.2 Schrödinger Equation

The Schorödigner equation is of the form:

$$i\hbar\frac{\partial}{\partial t}\Psi(q,t) = H\Psi(q,t)$$

It is an equation of time evolution of the wave function (or state vector). In this sense, the equation is also called the *time-dependent Schrödinger equation*, which can calculate the wave functions of the particles, given the potential in which they move.

In the *stationary state*, in which the energy $E$ is invariant over time, the following holds.

$$\Psi(q,t) = \exp(-iEt/\hbar)\Phi(q)$$

Substituting it for the above Schrödinger equation yields:

$$H\Phi(q) = E\Phi(q)$$

which is called the *time-independent Schrödinger equation*. It is nothing but the eigenvalue problem which is the basis for matrix mechanics as will be discussed in the next section. Notice that the solutions to the time-independent Schrödinger equation are simply the amplitudes of the solutions to the time-dependent Schrödinger equation.

Schrödinger equation is a differential equation which represents a wave function. It can explain all behaviors of light and matter. Namely, it can justify the wave-particle duality. Consequently, Schrödinger equation is a starting point of quantum mechanics.

## 3.3  Matrix Mechanics

Heisenberg established an alternative foundation for quantum mechanics as matrix mechanics. As noted above, it is known that matrix mechanics and wave mechanics are mathematically equivalent. Matrix mechanics is a theory founded on the Hilbert space.

**Werner Heisenberg (Germany: 1901-1976)**

Werner Heisenberg was born in Würzburg, Germany on 1901. From 1920 to 1923, he studied physics and mathematics at the University of Munich and University of Göttingen. In 1923, he took Ph.D. at the University of Munich. In 1927, he became a Professor at the University of Leipzig. In 1925, he proposed matrix mechanics as a foundation for quantum mechanics which was later shown to be equivalent to wave mechanics. He also published the Uncertainty Principle in 1927. He was awarded the Nobel Prize in Physics in 1932 for the creation of quantum mechanics. Heisenberg died in Munich, West Germany on 1976.

### 3.3.1  Linear Algebra

Now, we explain the basic notions in *linear algebra* in order to understand matrix mechanics; see Strang [77]. However, our exposition is rather abstract in that we start with a Hilbert space.

A *Hilbert space* is a linear vector space over a scalar field $\mathscr{C}$, designated by $\mathscr{H}$; see Young [87]. Let $a, b, c$ be vectors in the Hilbert space. Then, the following properties hold:

---

**Properties of Hilbert Space**

(1) For vectors $a$ and $b$, either $a = b$ or $a \neq b$.

(2) If $a = b$ then $b = a$.

(3) If $a = b$ and $b = c$, then $a = c$.

(4) $a$ and $b$ has the addition $a + b$. The addition $+$ satisfies the commutative law $a + b = b + a$ and the associative law $a + (b + c) = (a + b) + c$.

(5) If $\alpha$ is arbitrary scalar (real or complex), scalar multiplication $\alpha a$ is also a vector.

(6) The distributive law $\alpha(a + b) = \alpha a + \alpha b$ and $(\alpha + \beta)a = \alpha a + \beta a$ for scalar product $\alpha$ and $\beta$.

(7) $1a = a$ and $0a = 0$ hold. Here, 0 denotes the zero vector.

(8) For vectors $a$ and $b$, the scalar product $(a, b)$ is defined.

---

The space satisfying (1)-(7) is called the *vector space*, and the one satisfying (1)-(8) is called the Hilbert space.

The *scalar product* is also called the *inner product*. For a complex number $a + bi$, the *complex conjugate* is $a - ib$. We denote the complex conjugate of $(a, b)$ by $(a, b)^*$.

Scalar product is either real number or complex number, and satisfies the following properties.

---

(1)  $(a,a) \geq 0$ holds. If $a = 0$, then $(a,a) = 0$.

(2)  $(a, b+c) = (a,b) + (a,c)$ and $(a, \alpha b) = \alpha(a,b)$ hold.

(3)  $(a,b) = (b,a)^*$.

---

A tuple of finite or denumerably infinite vectors $a^{(1)}, a^{(2)}, \ldots$ is called *complete*, if arbitrary vector in the same space can be written as the form:

$$a = \sum_j c_j a^{(j)}$$

where $c_j$ denotes a number.

A tuple of vectors $a^{(1)}, a^{(2)}, \ldots$ is *independent*, if the formula for arbitrary vector in the same space,

$$\sum_j c_j a^{(j)} = 0$$

holds for every $c_j = 0$. Zero vector is not an element of independent tuples.

For a vector $a$, the *norm* is defined as follows:

$$\|a\| = \sqrt{(a,a)}$$

The norm is also called the *length*. If $\|a\| = 1$, then $a$ is called the *normal vector*. A *normalization* of any vector $a$ except the zero vector 0 can be obtained by dividing it by $\|a\|$.

We say that two vectors $a$ and $b$ are *orthogonal* if $(a,b) = a^*b = 0$. A tuple of vectors forms the *normal orthogonal system* if all elements in the tuple are normalized and they are orthogonal each other.

If any vector $a$ is transformed into another vector $Aa$, then $A$ is called the *linear operator*. Let $a$ and $b$ be vectors and $c$ be a scalar. Then, the linear operator $A$ satisfies the following properties.

---

(1)  If $a = b$, then $Aa = Ab$.

(2)  $A(ca) = c(Aa)$.

(3)  $A(a+b) = Aa + Ab$.

---

$A$ is also called the *linear transformation*.

Let $A$ and $B$ be linear operators, $a$ be a vector, and $c$ be a scalar. Then, we can define the following operators.

---

- $cA$ is the operator defined by $(cA)a = c(Aa)$.
- $A+B$ is the operator defined by $(A+B)a = Aa + Ba$.
- $AB$ is the operator defined by $(AB)a = A(Ba)$.
- $1$ is the operator defined by $1a = a$.

---

It is noticed that every linear operator can be described in terms of the corresponding *matrix*.

When $i \times j$ element in the matrix $A$ is $a_{ij}$, we write $A = [a_{ij}]$. If a linear operator is written as a matrix, the matrix multiplication evaluates a linear transformation.

Assume that independent vectors are no more than denumerably infinite in certain space, and that $\delta^{(1)}, \delta^{(2)}, \ldots$ be the tuple of normal orthogonal systems. Then, arbitrary vector $a$ can be written by the following form:

$$a = \sum_k a_k \delta^{(k)}$$

Here, $\delta^{(k)}$ is written as:

$$\sum_k \delta_k^{(j)} \delta^{(k)}$$

Note that the following holds.

$$\delta^{(j)} = \delta_{jk} = \begin{cases} 1 \ (j = k) \\ 0 \ (j \neq k) \end{cases}$$

Let $A$ be an arbitrary linear operator, and $A_{kl}$ be the element of a vector $A\delta^{(l)}$. Then, the element of $Aa$ is as follows:

$$(Aa)_k = \sum_l (a_l A\delta^{(l)})_k = \sum_l A_{kl} a_l$$

Since the tuple of whole elements $[A_{kl}]$ forms a matrix, the linear operator $A$ can be expressed as the matrix. The linear operator 1 is the *unit matrix*, and it can be written as the diagonal matrix $[\delta_{kl}]$.

For matrix $A$, vector $a$, and scalar $\lambda$, if

$$Aa = \lambda a$$

is satisfied, $\lambda$ is called the *eigenvalue* of $A$. The vector which is not the zero vector and corresponds to certain eigenvalue is called the *eigenvector*.

A *characterization equation* of the matrix $A$ is defined as:

$$\phi(\lambda) = |\lambda E - A| = 0$$

where $E$ denotes the unit matrix and $|A|$ the *determinant* of $A$. An eigenvalue is the solution of a characterization equation, and may be a complex number.

If matrices $A$ and $B$ have common eigenvector $a$, then we have $(AB - BA) = 0$. From this, $AB = BA$ holds. In this case, we say that $A$ and $B$ are *commutable*. Note that two matrices are not always commutable, giving the uncertainty principle.

There are several operators which are often used in matrix mechanics. The operator $P$ satisfying $P^2 = P$ is called the *projection operator*. The eigenvalue of a projection operator is either 0 or 1.

For any vectors $a$ and $b$, if $(Aa)^* b = a^* (A^* b)$, then the linear operator $A^*$ is called the *hermitian conjugate* of $A$. We say that the operator $A$ is an *hermitian operator* if

$A^* = A$. The eigenvalues of a hermitian operator are also real numbers. It should be noted that different eigenvalues of the hermitian operator are orthogonal each other. Because orthogonal vectors are independent each other, the eigenvector system of a hermitian matrix is a complete system.

We say that the operator $U$ is a *unitary operator* if $U^*U = U^*U = 1$, where $U^*$ is a hermitian conjugate of $U$. All the absolute values of the eigenvalues of a unitary operator are 1. If $Ua = \lambda a$, then we have that $\lambda^*\lambda = 1$ from

$$(Ua)^*(Ua) = a^*(U^*Ua) = a^*a = \lambda^*\lambda a^*a$$

Thus, the unitary transformation is the transformation which is invariant over the length of a vector.

### 3.3.2 Measurement

Now, we are ready to present a formalization of matrix mechanics. In quantum mechanics, the physical values like mass and position cannot be determined as a function. Instead, it can be determined probabilistically by means of the notion called the *observable*. It is obvious that the observable is closely related to the notion of *measurement*.

In general, the following two assumptions about measurement are accepted in quantum mechanics.

---

(1) We cannot infer the result of measurement $a$, but can infer whether it belongs to $\{a^{(ij)}\}$.

(2) We can infer the probability $p^j$ of the result of measurement $a^j$.

---

Here, the numbers $a^{(j)}$ represent the properties of measured physical system. The probabilities $p^{(j)}$ describe the properties of a *state* of the physical system.

The result of a measurement can be identified with real eigenvalues $a^{(1)}, a^{(2)}, \ldots$ of the hermitian operator $A$. Here, the operator $A$ denotes the measured physical value.

If the hermitian operator denotes measurable physical value, then it is called the *observable*. In other words, observables are hermitian operators. And the value of the observable is equal to an eigenvalue of its hermitian operator.

For example, in classical mechanics, the energy $E$ of a single particle system can be written as follows:

$$E = \frac{p^2}{2m} + V$$

But, in quantum mechanics, it can be described as the operator, i.e., observable, of the form:

$$H = -\frac{\hbar^2}{2m}\nabla + V$$

Next, we turn to the details of Heisenberg equation. As described above, in 1925, Planck and Einstein showed that the energy of radiation $E$ can be written as follows:

$$E = \hbar\omega$$

where $\omega$ is an angular frequency and $\hbar$ is the Planck constant. As a consequence, the following relation holds:

$$E^{(i)} - E^{(f)} = \hbar\omega$$

where $E^{(i)}$ is the energy of the initial state of an atom that radiates with the frequency of vibration $\omega$ and $E^{(f)}$ is the energy of the final state.

Let $a^{(i)}$ and $a^{(f)}$ be the eigenvector of $\lambda^{(i)}$ and $\lambda^{(f)}$ of the atom's energy $H$. Then, we can write in the following way:

$$Ha^{(i)} = \lambda^{(i)}a^{(i)}$$
$$Ha^{(f)} = \lambda^{(f)}a^{(f)}$$

From the above equation of the difference of energy of the initial and final states, and the operator of $A$ representing the physical value,

$$a^{(f)^*}(AH - HA)a^{(i)} = (E^{(i)} - E^{(f)})a^{(f)^*}Aa^{(i)}$$
$$= \hbar\omega a^{(f)^*}Aa^{(i)}$$

is derivable. Here, the left side describes the mean of commutable relation and the right side the mean of measurements of $A$, respectively.

Heisenberg assumed that the elements $a^{(f)^*}Aa^{(i)}$ of any operator $A$ change, like harmonic oscillators, with the frequency equal to the one of radiation as time passes. In other words, he assumed the following:

$$i\frac{d}{dt}(a^{(f)^*}Aa^{(i)}) = \omega a^{(f)^*}Aa^{(i)}$$

From this assumption, we have:

$$a^{(f)^*}(AH - HA)a^{(i)} = i\hbar\frac{d}{dt}(a^{(f)^*}Aa^{(i)})$$

If we assume that the vectors $a^{(i)}$ and $a^{(f)}$ are independent of time, we have the following equation which is called the *Heisenberg equation*:

$$AH - HA = i\hbar\frac{dA}{dt}$$

It is noticed that in the Heisenberg equation the operator $A$ is assumed to be variant over times. If we assume that the vectors $a^{(i)}$ and $a^{(f)}$ change over times, we obtain the equation of the form:

$$i\hbar\frac{da^{(i)}}{dt} = Ha^{(i)}$$

which is the Schrödinger equation. From the discussion, we can see that Heisenberg equation and Schrödinger equation, though they have different mathematical descriptions, are essentially equivalent. It follows that wave mechanics and matrix mechanics are equivalent theories.

The theory of matrix mechanics can give a precise physical interpretation of the physical system for the micro world. Note here that the interpretation is closely related to the principle of quantum computers. We can summarize the interpretation as follows.

First, the observable (hermitian operator) represents measurable physical value and its eigenvalues, and the results of measurement. If two different physical values can be simultaneously measured, then the observables representing these values are commutable. Otherwise the observables are not commutable.

Second, the normalized vectors describe the states of a physical system (e.g. system of an atom and system of particles). The projection of $a$ on the normalized eigenvectors $a^{(j)}$ of the observable $A$ represents the possibility of the eigenvalue $a^{(j)}$ obtainable for the measurement of physical values for $A$. In the state $a$, the probability that the eigenvalue $a^{(j)}$ of $A$ is measured is $c_j^* c_j$. Here, $c_j a^{(j)}$ is the projection of $a$ on the normalized eigenvector $a^{(j)}$.

Third, the mean (i.e., expected value) of measurements can be described as $a^* A a$, which is a real number. It can be always computed, and it is the basis for a quantum mechanical interpretation of measurement.

We can concisely state the interpretation. It is possible to describe all observables by matrices whose elements have indexes with two different energy levels. The set of eigenvalues of the matrix corresponds to the set of all possible values that observables can have. As described above, the eigenvalues are real due to the fact that matrices Heisenberg used are hermitian. After the measurement of an observable, the result is a certain eigenvalue whose corresponding eigenvector represents the state.

In quantum mechanics, we use the notion of *quantum number* which can characterize various quantum states in matrix (and wave) mechanics. There are four quantum numbers, i.e., the principal quantum number $n$, the orbital quantum number $l$, the magnetic quantum number $m$, and the spin $s$.

The *principal quantum number* determines the energy level, designating the principal electron shell. Its values include one and beyond, i.e., $n = 1, 2, 3, \ldots$.

The *orbital quantum number* determines the shape of the orbital of a quantum. Its value is either zero or a positive number, but is less than $n$, i.e., $l = 0, 1, 2, 3, \ldots, n-1$.

The *magnetic quantum number* determines the number of orbital's orientation within a subshell. Its value is between $-l$ and $l$, given a certain $l$, i.e., $m = -l, (-l+1), \ldots, -2, -1, 0, 1, 2, \ldots, (l-1), +l$.

The *spin* is a quantum number which designates the angular momentum intrinsic to quanta. Its value is a discrete value. The spin does not depend on other quantum numbers.

Here, we go into details of spin. Mathematically, a spin is a non-negative number $s$ satisfying $s(s+1) = \hbar^2 = \lambda$, if the eigenvalue of $S^2$ is $\lambda$.

Particles which have half-integer spin are called the *Fermion*. They include electron, proton and neutron. Particles which have integer spin are called the *Boson*. They include photon and meson.

The spin of electron is either $+\frac{1}{2}$ or $-\frac{1}{2}$. It shows the direction of the electron spin. If $s$ is positive, then the electron has an upward spin, i.e., spin up, and if $s$ is negative, then the electron has an downward spin, i.e., spin down.

The spin of a particle is described by a space vector $S$ which satisfies the following:

$$S \times S = i\hbar S$$

where $\times$ stands for *exterior product*. If we set $c = a \times b$, then the elements of $c$ are as follows:

$$c_1 = a_2 b_3 - a_3 b_2$$
$$c_2 = a_3 b_1 - a_1 b_3$$
$$c_3 = a_1 b_2 - a_2 b_1$$

Here, we prove that the above relation hold for an electron whose spin is $\frac{1}{2}$. For so doing, we need the *Pauli matrix*. There are three Pauli matrices, i.e., $\sigma_1, \sigma_2, \sigma_3$, which are defined as follows:

$$\sigma_1 = \begin{pmatrix} 0 & 1 \\ 1 & 0 \end{pmatrix}$$

$$\sigma_2 = \begin{pmatrix} 0 & -i \\ i & 0 \end{pmatrix}$$

$$\sigma_3 = \begin{pmatrix} 1 & 0 \\ 0 & -1 \end{pmatrix}$$

Pauli matrix is the hermitian matrix as well the unitary matrix. We may be able to define $\sigma_0 = 1$. Note also that the following hold.

$$\sigma_1 \sigma_2 = i\sigma_3$$
$$\sigma_2 \sigma_3 = i\sigma_1$$
$$\sigma_3 \sigma_1 = i\sigma_2$$
$$\sigma_i \sigma_j = -\sigma_j \sigma_i$$

Here, $i, j = 1, 2, 3, i \neq j$.

If we set:

$$S = \frac{1}{2}\hbar\sigma$$

Then the following holds.

$$S_1 S_2 - S_2 S_1 = \frac{1}{2}\hbar^2(\sigma_1\sigma_2 - \sigma_2\sigma_1)$$

$$= \frac{1}{4}\hbar^2(i\sigma_3 + i\sigma_3)$$

$$= \frac{1}{2}\hbar^2\sigma_3$$

$$= i\hbar S_3$$

Similarly, $S_2 S_3 - S_3 S_2 = i\hbar S_1$ and $S_3 S_1 - S_1 S_3 = i\hbar S_2$ follow. Consequently, the above relation $S \times S = i\hbar S$ is justified.

Since Pauli matrices $\sigma_1, \sigma_2$ and $\sigma_3$ have the relationship $\sigma^2 = \sigma_1^2 + \sigma_2^2 + \sigma_3^2 = 3$, we can describe as follows:

$$S^2 = \frac{3}{4}\hbar^2 = \frac{1}{2}\left(\frac{1}{2}+1\right)\hbar^2$$

which means that the spin of electron is $\frac{1}{2}$.

In 1927, Pauli introduced Pauli matrices to give a mathematical treatment of the spin of electron; see Pauli [65]. Observe that the work influenced Dirac to discover the Dirac equation. Nowadays, Pauli matrix is also applied to quantum gates which are the basis for quantum computers.

### 3.3.3  Bra-Ket Notation

Dirac devised a notation to serve as a mathematical foundation for quantum mechanics. It is called the *bra-ket notation* or *Dirac notation*; see Dirac [29]. By using the bra-ket notation, we can simplify the formalization of quantum mechanics.

The bra-ket notation is widespread in quantum mechanics. Notice that most phenomena in quantum mechanics can be described by means of it. Because it also becomes a standard notation for quantum computing, we review it below.

A *bra-ket* is composed of angle brackets and vertical bars, denoting the *inner product*, designated as $(a,b)$. We write it as follows:

$$(a,b) = \langle a|b\rangle$$

Here, $\langle a|$ is called the *bra* and $|b\rangle$ the *ket*, respectively. Although they are vectors, they are not written in bold.

Bra and ket can be used separately. If $a = (a_0, a_1, \cdots)$, then we have:

$$\langle a| = (a_0{}^*, a_1{}^*, \cdots)$$
$$|a\rangle = (a_0, a_1, \cdots)^t$$

Here, $^t$ denotes the transposition and $^*$ the complex conjugate. Thus, we can see that the bra is a row vector and the ket is a column vector.

The probability of the state $|a\rangle$ jumping to the eigenstate $|b\rangle$ can be expressed in the bra-ket notation as follows:

$$|\langle a|b\rangle|^2$$

It is possible to rewrite several concepts concerning the Hilbert space in terms of the bra-ket notation. The Hilbert space is the complete vector space with inner product and norm. If we deal only with the Hilbert space with a finite dimension, we have that $\mathscr{H} = C^n$, where $C$ denotes the set of complex numbers.

A vector $a\rangle \in \mathscr{H}$ is a normalized vector when $\||a\rangle\| = 1$. A denumerably infinite set of normalized vectors $B = \{|b_0\rangle, |b_1\rangle \cdots\}$ forms a normal orthogonal system if $\langle b_i | b_j \rangle = \delta_{ij}$.

If the normal orthogonal system $B$ is written as

$$|a\rangle = \sum_i \lambda_i |a_i\rangle$$

then it is called the *basis*.

Now, we introduce *tensor product*, denoted $\otimes$, into the Hilbert space. $a \otimes b$ can be expressed in the bra-ket notation as follows:

$$|b\rangle\langle a|$$

Let $\mathscr{H}_\infty$ and $\mathscr{H}_\in$ be the Hilbert systems with the basis $B_1$ and $B_2$. Then, we have:

$$\mathscr{H} = \mathscr{H}_\infty \otimes \mathscr{H}_\in = \left\{ \sum_{|i\rangle \in B_1} \sum_{|j\rangle \in B_2} c_{ij} |i,j : c_{ij} \in C \right\}$$

where $\mathscr{H}$ is the Hilbert space satisfying that $B = B_1 \times B_2$ and $\langle i,j | i',j' \rangle = \langle i | i' \rangle \langle j | j' \rangle = \delta_{ii'}\delta_{jj'}$ ($|i\rangle, |i'\rangle \in B_1 | j\rangle, |j'\rangle \in B_2$).

If $A, B, C, D, U$ are matrices, $a, b, c$ are vectors, and $\alpha, \beta$ are scalars, then the following relations hold.

$$(A \otimes B)(C \otimes D) = (AC \otimes BD)$$
$$(A \otimes B)(a \otimes b) = (Aa \otimes Bb)$$
$$(a + b) \otimes c = a \otimes c + b \otimes c$$
$$\alpha a \otimes \beta b = \alpha\beta(a \otimes b)$$

$$\begin{pmatrix} A & B \\ C & D \end{pmatrix} \otimes U = \begin{pmatrix} A \otimes U & B \otimes U \\ C \otimes U & D \otimes U \end{pmatrix}$$

$$(A \otimes B)^* = A^* \otimes B^*$$

where $A^*$ is the conjugate transposition of $A$.

A linear operator in $C^n$ can be defined as a $n \times n$ matrix in which $a_{ij} = \langle i | A | j \rangle$, i.e.,

$$A = (a_{ij}) = \sum_{i,j} a_{ij} |i\rangle\langle j|$$

Now, we describe an eigenvalue in the bra-ket notation. First, the normal equation is written as follows:

$$A |a\rangle = \lambda |a\rangle$$

A eigenvalue is its solution $\lambda$ and the eigenvector for the eigenvalue $\lambda$ is $|a\rangle$.

Bra-ket notation enables us to simplify the description of basic equations in quantum mechanics. For instance, we can write the Schrödinger equation in the following way:

$$i\hbar \frac{\partial}{\partial t} |a\rangle = H |a\rangle$$

## 3.4 Uncertainty Principle

Although quantum mechanics can deal with phenomena in the micro world, Heisenberg's *uncertainty principle* holds for the measurement of phenomena; see Heisenberg [48]. As explained later, the uncertainty principle can be also applied to quantum computing.

According to the uncertainty principle, the position and momentum of a particle cannot be simultaneously measured with arbitrarily high precision. There is a minimum for the product of the uncertainties of these two measurements.

### 3.4.1 Starting Point

Let measurement errors of physical value $p$ and $x$ be $\nabla p$ and $\nabla x$, respectively. Then, the uncertainty principle can be expressed as follows:

$$\nabla p \nabla x \approx \hbar$$

which means that two physical values cannot be simultaneously determined precisely. In other words, the uncertainty principle says that the combination of the error in position times the error in momentum must always be greater than Planck's constant.

So, you can measure the position of an electron to some accuracy, but then its momentum will be inside a very large range of values. Note also that you can measure the momentum precisely, but that its position is unknown.

For instance, we cannot simultaneously determine the position $x$ of a particle and its momentum $p$. In other words, as we attempt to measure either the position and the momentum precisely, the measurement of the other becomes imprecisely. But this is odd in classical mechanics.

### 3.4.2 Mathematical Formulation

Now, we look at the uncertainty principle mathematically. Let $A$ and $B$ be observables, $a$ and $b$ be the corresponding eigenvalues, $|\psi\rangle$ be a state before the measurement was performed.

Then, if the state $|\psi\rangle$ after the measurement is defined by the eigenvectors of $A$ and $B$, we have:

$$A \mid \psi' \rangle = a \mid \psi' \rangle$$
$$B \mid \psi' \rangle = b \mid \psi' \rangle$$

This is equivalent to the case that it is commutable, i.e., $[A,B] = AB - BA = 0$.

Let the measurement of an observable $A$ be $\mid \psi \rangle$. Then, its expected value (i.e., mean) is expressed as follows:

$$\langle A \rangle = \langle \psi \mid M \mid \psi \rangle$$

Here, if $\delta A = A - \langle A \rangle$, then we have:

$$(\nabla A)^2 (\nabla B)^2 = \langle (\delta A)^2 \rangle \langle (\delta B)^2 \rangle$$
$$= \langle \psi \mid (\delta A)(\delta A) \mid \psi \rangle \langle \psi \mid (\delta B)(\delta B) \mid \psi \rangle$$

Since $\delta A$ and $\delta B$ are hermitian,

$$(\nabla A)^2 (\nabla B)^2 = \parallel \delta A \mid \psi \rangle \parallel^2 \parallel \delta B \mid \psi \rangle \parallel^2$$

follows. From Schwarz's inequality

$$\mid \langle \psi \mid \phi \rangle \mid \leq \parallel \psi \parallel \parallel \phi \parallel$$

and $[A,B] = [\delta A, \delta B]$,

$$(\nabla A)(\nabla B) \geq \frac{1}{2} \mid \langle [A,B] \rangle \mid$$

can be obtained, and the left side of the formula is equivalent to $\frac{1}{2}\hbar$.

The intuition of the above can be stated as follows. Suppose that two observables $A$ and $B$ are measured simultaneously. Then, the state of the system should collapse to a common eigenvector of the two observables.

Eigenvectors in most matrices are not equal in general. Thus, the eigenvectors of $A$ and $B$ can be assumed to be not equal. This means that observables $A$ and $B$ cannot be simultaneously determined precisely.

The uncertainty principle claims that the measurement of physical values cannot be determined and that it can be only probabilistically described. It is well known that there are several interpretations of the uncertainty principle, none of which can be regarded conclusive.

The interpretation of the uncertainty principle is philosophically interesting, and is closely related to the called the *measurement problem*, which has been defined in various ways. However, we can simply define it as the problem of the connection between a quantum system and reality.

Thus, the measurement problem involves various measurements. The problem has been originally discussed as the problem of how wave function collapse occurs in the context of the Copenhagen interpretation. We will discuss the Copenhagen interpretation later.

The problem has been also discussed as the problems of the uncertainty principle, superposition states, etc. Thus, the measurement problem is now regarded as one of the open problems in the philosophical interpretation of quantum mechanics. We believe that the solution of the problem is also helpful to the development of quantum computing.

## 3.5 Quantum Systems

Von Neumann proposed a mathematical foundation for quantum mechanics based on the Hilbert space; see von Neumann [82].

**John von Neumann (Hungary: 1903-1957)**

John von Neumann was born in Budapest, Hungary on 1903. He entered the University of Berlin to study chemistry in 1921. He received his doctorate in mathematics from the University of Budapest with a thesis on set theory in 1926. He contributed to many areas from mathematics to quantum mechanics, computer science, game theory, and so on. In 1932, he published a theoretical foundation for quantum mechanics. In 1936, he proposed quantum logic, with Birkhoff. In 1944, he provided a mathematical foundation for game theory, with Morgenstern. In 1945, he proposed the architecture of computers, now known as the Neumann-type computer. In the early 1950's, he designed a cellular automaton implementing a universal self-replicating structure. Von Neumann died in Washington, USA on 1957.

### 3.5.1 Postulates of Quantum Systems

Von Neumann formulated a *quantum system* axiomatically by specifying several postulates. In this sense, quantum computers can be seen as one of the quantum systems. It is thus interesting to see his postulates in connection with quantum computers. We check these postulates in some detail.

Postulate 1 is concerned with a quantum system and a *state*.

---

**Postulate 1**

A quantum system $S$ is described by the Hilbert space $\mathscr{H}$.

---

A state is expressed as a *state vector* which is a normalized vector $|\psi\rangle$. A state vector represents physical states of a physical system. One of the simplest quantum systems is a *qubit* (quantum bit) whose state space $\mathscr{B}$ is $C^2$. A state of qubit $|\psi\rangle$ is represented as the linear combination of basis states $|0\rangle$ and $|1\rangle$, i.e.,

$$|\psi\rangle = \alpha|0\rangle + \beta|1\rangle$$

where $\alpha, \beta \in C$ and $|\alpha|^2 + |\beta|^2 = 1$.

This representation is called the *superposition*, whose state is specified both by $|0\rangle$ and $|1\rangle$. A qubit is a basic unit of information in quantum computing. Namely, the role played by a qubit in quantum computers is the same as the one by a bit in classical computers. However, we will explain details of qubit in Chapter 4.

Postulate 2 is the assumption about the *evolution* of a state.

---

**Postulate 2**

The time evolution of states in a quantum system is described by means of the Schrödinger equation.

---

As described above, the Schrödinger equation, which is one of the basic equations in quantum mechanics, can be written as follows:

$$i\hbar\frac{\partial}{\partial t}|\psi\rangle = H|\psi\rangle$$

In general, we assume that $\hbar = 1$ for the measurement in a quantum system. The Hamiltonian operator $H$ can completely formulate the mechanics of a quantum system in question.

Postulate 3 is the assumption about the measurement.

---

**Postulate 3**

An observable is described by the Hamiltonian operator. If the measurement of the observable $A$ is performed and the eigenvaules of $A$ can be computed, then the result of the measurement can be probabilistically described by means of the eigenvalues.

---

Although an observed value is described by the observable $M$, we can write it by spectrum decomposition in the following way.

$$M = \sum_m mP_m$$

Here, $P_m$ denotes the *projection operator* to the eigenspace which corresponds to the eigenvalue $m$.

The eigenvalue $m$ of $M$ denotes possible results of the measurement. Namely, the result $m$ is obtainable from the measurement $|\psi\rangle$. Let the state before the measurement be $|\psi\rangle$. Then, the probability $p(m)$ of result of the measurement is given by the value $\langle\psi|P_m|\psi\rangle$.

Consequently, the state $|\psi'\rangle$ after the measurement is expressed as follows:

$$|\psi'\rangle = \frac{1}{\sqrt{p(m)}}P_m|\psi\rangle$$

It is noticed that the uncertainty holds for measurements in a quantum system.

Postulate 4 is the assumption which concerns the composition of quantum systems.

---

**Postulate 4**

The state space $H$ of the composite quantum system of $n$ quantum systems with the state spaces $H_1, ..., H_n$, state vectors $| \psi_1 \rangle, ..., | \psi_n \rangle$ becomes tensor product $H = H_1 \otimes \cdots \otimes H_n$ of state spaces of subsystems and tensor product $| \psi \rangle = | \psi_1 \rangle \otimes \cdots \otimes | \psi_n \rangle$ of state vectors of subsystems.

---

Not all state vectors $H$ can be defined by the tensor product of the state vectors of $H_1, ..., H_n$. The state which cannot be described by tensor product is called the *entangled state* and it applied to quantum teleportation.

## 3.5.2   Some Foundational Problems

Finally, we discuss several problms on the foundations for quantum mechanics. They are all related to the measurement problem. One of the well known interpretations of measurement is the *Copenhagen interpretation* due to Bohr.

**Niels Bohr (Denmark: 1885-1962)**

Niels Bohr was born in Copenhagen, Denmark on 1885. In 1903, he entered Copenhagen University to study mathematics and philosophy. In 1911, he received his doctorate from the university. In 1916, he became Professor of Theoretical Physics at Copenhagen University. He proposed a model of the structure of atoms in 1933, inspired by Rutherford's discovery of the atomic nucleus. He received the Nobel Prize in Physics in 1922 for his work on the structure of atoms. He also worked on the Manhattan Project during the World War II. Bohr died in Copenhagen, Denmark on 1962.

## Copenhagen Interpretation

The Copenhagen interpretation is the principle to explain the wave-particle duality in terms of a *collapse* of the wave function defended by Bohr. It is based on three basic ideas.

First, a wave/particle can be completely described by the wave function. This means that any information that cannot be derived from the wave function does not exist.

Second, when a measurement of the wave/particle is performed, its wave function collapses. Even if we precisely measure the momentum of a particle, its wave function suddenly changes from a wave made up of many momenta, to a wave with only one momentum. This is called the *collapse*.

Third, if two properties are related by an uncertainty relation, then we cannot make a measurement which can simultaneously determine both properties to a precision greater than the uncertainty relation allows. Consequently, when we measure the position of a particle, we affect its momentum, and vice versa.

For example, according to the Copenhagen interpretation, when the measurement of an electron is done, the electron (known as electron cloud) which is the wave of superposition state collapses a particle.

In the Copenhagen interpretation, it can be understood that there exists an observable when a physical system is in the eigenstate of the corresponding operator. This implies that the result of physical experiments can be expressed in classical physics.

In practice, it is possible to determine the state of an electron cloud by the Schrödinger equation. If a measurement is done, the electron can be measured as the particle. In the interpretation, an electron exists somewhere in an electron cloud. When measuring, the electron cloud instantaneously collapses in a particle and it can be measured as an electron.

In this regard, we cannot measure an electron cloud itself. The Copenhagen interpretation is one of the promising interpretations of quantum mechanics. But, it is odd. In fact, it was refuted by Schrödinger who gave a paradox known as *Schrödinger's cat* in 1935.

## Schrödinger's Cat

Schrödinger's cat is a thought experiment, demonstrating the conflict between the behavior described by quantum mechanics in the microscopic level and the behavior described by classical mechanics in the macroscopic level.

The story is as follows. A cat is placed in a close box, together with radium, the Geiger counter and a close flask of poison gas (Fig. 3.1).

If radium emits an *alpha*-ray by $\alpha$-decay, then the Geiger counter detects $\alpha$-ray and wire cuts. Then, the hammer falls and hits the flask. From the flask, poison gas rises, killing the cat. There are several versions of Schrödinger's cat, the essential idea is equivalent.

Let the probability that $\alpha$-decay will happen within a hour be 50%. Then, in quantum mechanics, the state is interpreted as the superposition state in which both

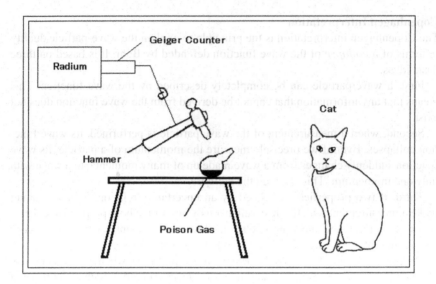

**Fig. 3.1** Schördinger's cat

the probability $\alpha$-decay will happen and will not are 50%. The life and death of the cat is finally determined by a measurement.

In the superposition state, whether $\alpha$-decay happens or not and the life and death of the cat are both uncertain. In other words, there simultaneously exist both alive and dead cats until a measurement is done. Schrödinger considered the situation as a paradox because there are no such cats in the real world.

The paradox led Schrödinger to question the concept of uncertainty itself in quantum mechanics. This is to say that the situation created by Schrödinger's cat is odd if we follow quantum mechanics. This is because the state of a cat which is simultaneously in the dead and alive states will be determined by instantly collapsing one of the states after by a measurement.

Thus, we can say that Schrödinger's raised the defects of the Copenhagen interpretation. Actually, the paradox is regarded as one of the important problems in the measurement problem. However, currently we have no reasonable interpretations of Schrödinger's cat.

It is a well known fact that Einstein was also critical for foundations for quantum mechanics as he showed the EPR paradox. The fact is very interesting. In fact, he thought that quantum mechanics cannot provide a complete description of physical systems.

**EPR Paradox**

In 1935, Einstein, Podolsky and Rosen presented the so-called *EPR paradox*. It was given as the paradox to demonstrate the shortcoming of the formulation of quantum mechanics; see Einstein, Podolsky and Rosen [33].

Now, we explain the details of the EPR paradox. Consider two particles which move in opposite direction after a collision. Here, we know the total momentum $p = p_1 + p_2$. But the momentum $p_1$ and $p_2$ are individually uncertain. Notice also that the position $q_1$ and $q_2$ are individually uncertain although the position $r = q_1 - q_2$ is known.

In this situation, the following facts hold. If the momentum $p_1$ of one particle is measured, then the momentum $p_2$ of the other particle is determined. But, if the position $q_1$ of one particle is measured, then the position $q_2$ of the other particle is determined. The situation in the EPR paradox in fact illustrates an *entangled state*, which plays a role in quantum computing.

According to Einstein, Podolsky and Rosen, these facts involve the following two paradoxical issues.

---

(1) Although the second particle is far apart from the position of a measurement, the measurement of physical values concerning the first particle affects that concerning the second particle.

(2) Although the position or the momentum of the second particle can be determined, quantum mechanics claims that they cannot be determined.

---

(1) is called the *non-local correlation*, but it is inconsistent with the principle of constancy of light velocity in Einstein's Theory of Relativity. For this implies that the affect of the collapse of a wave transmits in the velocity faster than the light velocity. But, from the principle of light velocity, no velocity of particles cannot exceed that of light.

(2) says that the measurement of a quantum system is essentially uncertain but that the physical values in question are determined. Namely, physically correct theories follow determinism, but quantum mechanics only gives probabilistic description.

They claimed that these points are odd, and later their claims were called the EPR paradox. (1) suggests the so-called *hidden variable theory*. The influence like (1), called the *EPR correlation*, was formalized as *Bell's inequality* by Bell in 1964; see [5].

Bell proved that there exists no hidden variable theory by means of Bell's inequality. In other words, we have phenomena which cannot be formalized by the hidden variable theory.

The EPR correlation is experimentally verified, and it is believed that there is a non-locality which Einstein disliked. Non-locality can be interpreted as the entangled state in quantum mechanics. The concept can be applied to quantum code and quantum teleportation as will be discussed later.

(2) can be resolved by considering the fact that the devices needed for the measurement of momentum are different from those needed for the measurement of position and that these measurements are not compatible.

After all, it can be interpreted that the EPR paradox is based on the misunderstanding in classical physics that physical values have determined values even if they are not measured. However, we can say that the proposal of the EPR paradox addresses some important ideas in the present quantum computing like the entangled state.

Schrödinger's cat and the EPR paradox challenge the current foundations for quantum mechanics. Unfortunately, we can know no viable solutions to these paradoxes. We believed that the solutions will be able to address some aspects of quantum computing as well as the philosophy of quantum mechanics.

# Chapter 4
# Quantum Computers

**Abstract.** Chapter 4 explains quantum computer in some details. After the historical survey of the origins of quantum computers, we introduce the quantum Turing Machine. Qubit, quantum gates and Shor's algorithm are also discussed.

## 4.1 Origins of Quantum Computers

Here, we survey the origins of quantum computers. First, we introduce Feynman's ideas of the starting point of quantum computers. As described in Chapter 1, the ideas of quantum computers have been presented by Feynman in his lecture held in MIT on 1981; see Feynman [36].

Feynman took up several phenomena in quantum mechanics by means of classical computers, and discussed some problems in their efficient simulations. Based on the discussions, he convinced that quantum systems can be efficiently simulated by the computers by making use of quantum-mechanical effects. In fact, Feynman's paper is a pioneer work on quantum computing.

**Richard Feynman (USA: 1918-1988)**

S. Akama, *Elements of Quantum Computing,*
DOI: 10.1007/978-3-319-08284-4_4, © Springer International Publishing Switzerland 2015

Richard Feynman was born in Far Rockway, USA on 1918. He received B.Sc. from
MIT in 1939 and Ph.D. from Princeton University in 1942. He became Professor of
Theoretical Physics at Cornell University in 1945. He contributed to several fields in
physics, in particular, quantum mechanics and quantum electrodynamics. He received
the Nobel Prize in Physics in 1965, with Schwinger and Tomonaga for his fundamental work in quantum electrodynamics, with deep consequences for the physics of
elementary particles. He also opened the door of quantum computing. He wrote several textbooks including the Feynman Lectures on Physics. In Feynman died in Los
Angeles, USA on 1988.

### 4.1.1  Feynman's Ideas

Here, we summarize the point of Feynman's ideas. He started with the problem.
I.e., What kind of computer are we going to use to simulate physics? The question
can be restated as follows. Can physics be simulated by a universal computer, i.e.,
Turing Machine or Neumann-type computer?

Feynman pointed out that although natural laws are *reversible*, the rule for computers are not. Based on the fact, he said that we need a new computer, which is
different from Turing Machine and obeys the principles of quantum mechanics, i.e.,
quantum computer[1].

Reversible computation is the computation which satisfies the laws of thermodynamics. Thus, its computational process is reversible. There are two types of reversibility, i.e., physical reversibility and logical reversibility. Physically reversible
process is the process which does not increase entropy. Logically reversible process
is the process which can construct the inputs from the outputs.

In computer science, reversible computation has been already investigated. Landauer discussed the relationship between physically and logically reversible processes, showing that while all logically reversible processes may be accomplished
as a thermodynamically reversible process, some logically irreversible processes
cannot. Namely, it means that if the process is physically reversible then it must be
also logically reversible. Later, Bennett extensively studied the notion of reversible
computation; see Landauer [57], Bennett [7] and Fredkin and Toffoli [39].

As an important requirement of his discussion, Feynman mentioned the following
rule of simulation, i.e., the number of computer elements required to simulate a large
physical system is only to be proportional to the space-time volume of the physical
system. This means that he referred to simulations with exponential complexity as
being against his rule of simulation. Feynman introduced some examples and argued
that classical computers are not adequate to simulate physical phenomena.

The first example is a simulation of time. Feynman assumed that the time is discrete. As the computer goes from state to state like in cellular automata, he claimed
that the time is not simulated at all, it is imitated by means of the notion of transition of states in the computer. But he noted that classical physics is local, causal,
and reversible and therefore adaptable to computer simulation.

---

[1] Feynman seemed to prefer to use the term 'quantum-mechanical computer' rather than the
term 'quantum computer'.

He mentioned the following example in space time domain. The state $s_i$ at the space-time point $i$ is a given function $F_i(s_j, s_k, ...)$ of the state at the points $j, k$ in the some neighborhood of $i$:

$$S_i = F_i(s_j, s_k, ...) \tag{1}$$

If $F_i$ only contains the points previous in time, it is possible to do the computation in a classical way. However, if $F_i$ is a function of both future and past, it is not clear whether there is an algorithm for computing a solution. It should be also stressed that the computation may not be possible, even if the function $F_i$ is known.

The second example is concerned with simulating probability. Since in quantum mechanics, we can only infer the probability of phenomena, the simulation of probability is of special importance. According to Feynman, there are two different methods of simulating probability in a computer. One method is to compute a probability first and then to interpret it to describe the nature. The other method is to simulate a probability by using probabilistic computers.

However, the first method faces a problem with discreting probability. If there are only $k$ digits to represent the probability of an event, the computer is useless when the probability of something happening is less that $2^{-k}$. If we have $R$ particles, then we must describe the probability of a situation by the probability to these particles at points $x_1, x_2, ..., x_R$ at the time $t$. This implies that $k$-digit number would be needed to describe the state of the system for every arrangement of $R$ values of $x$. Thus, we need around $N^R$ configurations if there are $N$ points in space. This means that if the number of physical values increases, the computation of the size increases exponentially. Therefore, the method is inadequate for simulation.

The difficulty with the second method is that the transition of states is probabilistically, implying the nature is not predictable. This means that we cannot simulate the nature. In other words, it will not be exactly the same as nature, even if the probabilistic computer will imitate the nature. To overcome the problem, Feynman proposed a Monte Carlo method. If a particular type of experiment is repeated sufficient number of times, the corresponding probability could be found statistically. Feynman described the nature of probabilistic computer as *local* probabilistic computer, in which the behavior of one region can be determined by disregarding the other regions.

Feynman showed the equations of the following form. At each point $i = 1, 2, , ..., N$ in space, there is a state $s_i$ chosen from a small state set. Then, the probability to find some configuration $\{s_i\}$ is $P(\{s_i\})$. The state transition from a state at the time $t$ to the next state at the time $t + 1$ can be given as follows:

$$P_{t+1}(\{s_i\}) = \sum_{s'} \left( \prod m(s_i \mid s'_j, s'_k, ...) \right) P_t(\{s'\}) \tag{2}$$

where $m(s_i \mid s'_j, s'_k, ...)$ denotes the probability that we move to state $s_i$ at point $i$ when neighbors have values $s'_j, s'k, ...$ in the neighborhood of $i$.

Since $j$ moves far from $i$, $m$ is ever less sensitive to $s'_j$. At each change, the state at a particular point $i$ will move from what it was to a state $s$ with a probability $m$ that depends only on the state of the neighborhood. Then, a probability of transition can be obtained. The state transition is probabilistic rather than definite. This is also true for cellular automata. From the discussions on simulating time and probability reviewed above, even if we can simulate imitate the nature, we can see that the nature is not predictable.

### 4.1.2 Simulating Quantum Mechanics

Based on some arguments, Feynman concluded that for simulating quantum mechanics, we need a probabilistic computer or a new type computer. The latter corresponds to what we now call a quantum computer. It is to be noted that Feynman used the term *quantum simulator* as a model of quantum computers. But, his paper showed only a sketch of quantum computers.

Feynman's prospect was that if the machine based on the elements of quantum mechanics is construed then it will be able to simulate arbitrary quantum system probabilistically. His insight is regarded as the origin of current research on quantum computers. And he discussed some phenomena to address the usefulness of quantum simulators.

First, he insisted that quantum systems cannot be described by means of classical computers. The claim reveals that we cannot represent the results of quantum machines with a classical universal device due to the so-called the *hidden variable problem* which addresses that a physical system cannot completely describe the nature. If we try to put the quantum equations in a form as close as possible to classical equations, we cannot simulate the quantum equations in the normal way because there are too many variables. There are thus no Turing Machines which imitate with the same probability what nature does and output the same probabilities that we observe in quantum mechanical experiments.

Second, he said that the concept of negative probability is needed to simulate quantum mechanics. To justify the point, Feynman give the following example. For many interacting spins on a lattice, we can give a probability for correlated "probabilities" as follows.

$$F(s_1, s_2, ..., s_N) \tag{3}$$

where $s_i \in \{++, +-, -+, --\}$. Then, the quantum mechanical equation which tells us what the changes of $F$ are with time is of the form (4).

$$F_{t+1}(\{s_i\}) = \sum_{s'} \left( \prod M(s_i \mid s'_j, s'_k, ...) \right) F_t(\{s'\}) \tag{4}$$

Compare (2) with (4). We have $F$ instead of $P$.

$M(s_i \mid s'_j, s'_k), ...)$ could be interpreted as the "probability". The "probability" $M$ is a probability in the usual sense, since it may be negative. It is impossible to simulative *negative probability* in classical computers[2].

---

[2] The idea of negative probability in quantum mechanics has been already proposed by Dirac in 1942; see Dirac [30].

Third, he exemplified the *polarization of photons* discussed in the EPR paradox as two-states systems. The polarization of photons is the phenomenon that the direction of radiation of photon is biased. The polarization of the two photon system can be formalized as the entangled state in quantum mechanics. Here, the probability of the states of each photon cannot be computed locally. For two photons in the entangled state, the measurement of one photon can determine the measurement of the other photon. This is the same as the case in the EPR paradox discussed in the last chapter.

Finally, Feynman discussed the two photon correlation experiment, dealing with the correlation of two photons. In this connection, he argued that there is a possibility of resulting negative probability in formalizing a composite system based on local probabilities.

This is the outline of Feynman's discussion, the need of quantum computers was best expressed in the last of the paper:

> And I'm not happy with all the analyses that go with just the classical theory, because nature isn't classical, dammit, and if you want to make a simulation of nature, and if you want to make a simulation of nature, you'd better make it quantum mechanical, and by golly it's a wonderful problem, because it doesn't look so easy. Thank you.

Additionally, the description appears to foresee that the work on quantum computers is very difficult. Indeed Feynman's paper gave basic ideas of quantum computers, but it did not present a concrete computational model. We believe that in the period Feynman was not be able to work out the formalization of a quantum Turing Machine although he have the background on Turing Machine. However, the book (i.e., Feynman, Hey and Allen [37]), which was published after his death, includes more concrete theory of quantum computers; see Feynman [37].

In fact, if we read the book carefully, we can see that Feynman's theory of quantum computers is essentially equivalent to that of quantum gates presented below. Feynman used the term *quantum mechanical computer* instead of the terms like quantum computers or quantum simulator. We can also find the detailed treatment of quantum gates. Since Feynman is a physicist, his theory was naturally based on physical thinking. In fact, his interest seemed to concentrate on the problem of energy consumption rather than the method of constructing quantum computers. Therefore, it is doubtful whether Feynman's theory of quantum computers could be fully understood by many computer scientists in the period. Now, only the fact that Feynman's ideas led to quantum computers appears to be evaluated. It may be unfortunate that Feynman died before the research on quantum commuters was active. Feynman died in 1988.

## 4.2 Quantum Turing Machine

The notion of quantum computers has been suggested by Feynman as described in the previous section, but the details have been remained open. The discussion, which is more close to computer science, on quantum computers was done by English physicist Deutsch in 1985; see Deutsch [24].

**David Deutsch (UK: 1953-)**

David Deutsch was born in Haifa, Israel on 1953. He is one of the major figures in quantum computing. In 1985, he proposed a theory of quantum computer as the first computational model of quantum computers. He also published a theory of quantum gates in 1989.

Deutsch physically reformulated the *Church-Turing principle* (CTP), usually called the Church's Thesis, and proposed the *universal quantum computer*, which is quantum mechanical generalization of the universal Turing Machine. Deutsch's universal computer called *QC* is computationally equivalent to *quantum Turing Machine*. But, he did not provide a formal definition of quantum Turing Machine. This may be funny as a computational model for quantum computers. We do not know why Deutsch formalized a computational model of quantum computers as in the form of *QC*. However, we could say that Deutsch proposed quantum Turing Machine in the broad sense.

### 4.2.1   Church-Turing Hypothesis and Beyond

Now, we discuss Deutsch's approach in details. Although it is obvious that Deutsch was influenced by Feynman, he started with more abstract level, namely the analysis of Church's Thesis. Deutsch called it the *Church-Turing hypothesis* (CTH); see Church [19] and Turing [79]. CTH can be written as follows:

**Church-Turing Hypothesis (CTH)**
Every 'function which would naturally be regarded as computable' can be computed by the universal Turing Machine.

It could be simply paraphrased that every computable function can be computed by the universal Turing Machine. As described in Chapter 2, the function computable by the universal Turing Machine is Turing computable. CTH is not a theorem but a hypothesis. Thus, CTH cannot be mathematically proved, but can be assumed. Actually, in computer science, it has been recognized that it is true.

Supposing CTH, we can understand that all possible formalizations of the intuitive mathematical notion of 'algorithm' or 'computation' are equivalent to each other. Therefore, by CTH, it is considered that Turing Machine, recursive function and $\lambda$-calculus are equivalent theories of computation in view of CTH. Although CTH is the hypothesis in theory of computation, Deutsch thought that the corresponding hypothesis is also *physically* correct. He called it the *Church-Turing principle* (CTP) to distinguish it from CTH. Recently, CTP is also called the Church-Turing-Deutsch principle (CTDP). We can write CTP as follows:

**Church-Turing Principle (CTP)**
Every finitely realizable physical system can be perfectly simulated by a universal model computing machine operating by a finite means.

We need some words about the description. According to Deutsch, CTP is physical and unambiguous principle. In CTH, computable function is regarded as a function which can be *concretely* computed. But, the term *concretely* or 'would naturally be regarded as' is mathematically ambiguous. In contrast, in CTP, the term *perfect simulation* is used. It is more physical and means that the physical laws involve the laws of machine (computer). Observe here that the physical idea similar to the perfect simulation can be also found in Feynman's papers. In fact, Feynman consider it as simulating with physics with computers.

Deutsch presented a definition of perfect simulation. A computing machine $\mathcal{M}$ is capable of perfectly simulating a physical system $S$, under a given labelling of their inputs and outputs, if there exists a program $\pi(S)$ for $\mathcal{M}$ that renders $\mathcal{M}$ computationally equivalent to $S$ under the labelling. Here, the labelling denotes the physical representation of data. Quantum mechanically, the labelling is a set of ordered pairs consisting of Hamilton operator and its eigenvalues.

Thus, $\pi(S)$ for $\mathcal{M}$ converts $\mathcal{M}$ into 'black box' functionally indistinguishable from $S$. The 'finitely realizable physical systems' mentioned above must include any physical object upon which experimentation is possible. On the other hand, the 'universal computing machine' need only be an idealized (but theoretically permitted) finitely specifiable model. The term 'finite means' can be defined axiomatically, without restrictive assumptions about the form of physical laws. Thus, it can be interpreted as finite steps or time. CTP is obviously stronger than CTH in that CTP is a physical (quantum mechanically) hypothesis. It is so strong that it is *not* satisfied by the Turing Machine in classical physics.

But the reader might wonder why the computer based on difficult theory like quantum mechanics is needed. This is a natural question. The answer is, from Feynman's and Deutsch's discussion, that the Turing Machine based on classical physics is not suitable and that a new machine founded on quantum mechanics, i.e., quantum computer is needed.

In other words, a computer satisfying CTP is not a classical computer but a quantum computer. Deutsch says the following to motivate quantum computers in Deutsch [24].

Every existing general model of computation is effectively classical. That is, a full specification of its state at any instant is equivalent to the specification of a set of numbers, all of which are in principle measurable. Yet according to quantum theory there exist no physical systems with this property.

From the quotation, we can see that quantum computers have the principle which is different from classical computers. In this regard, Deutsch's starting point, i.e., CTP is conceptually defensible.

### 4.2.2  Universal Quantum Computer

After the detailed discussion on CTP, Deutsch presents a general quantum model of computation called the *universal quantum computer* denoted $Q$. It can perfectly simulate every finite, realizable physical system. It can simulate ideal closed (zero temperature) systems, including all other instances of quantum computers and quantum simulators, with arbitrarily high but not perfect accuracy. Therefore, upon CTP, $Q$ is a model equivalent to quantum mechanical version of Turing Machine, i.e., quantum Turing Machine. There is a sense in which Deutsch proposed a quantum Turing Machine, although he did not directly formalized a quantum Turing Machine.

Quantum Computer $Q$ has, like a Turing Machine, two components, i.e., a finite processor and an infinite memory, of which only a finite portion is every used. The computation proceeds in steps of fixed duration T, and during each step only the processor and a finite part of the memory interact, the rest of the memory remaining static.

The processor consists of $M$ two-state observables, i.e.,

$$\{\hat{n}_i\}  (i \in Z_M)$$

where $Z_M$ is the set of integers from 0 to $M - 1$. The memory consists of an infinite sequence of two-state observables, i.e.,

$$\{\hat{m}_i\}  (i \in Z)$$

which corresponds to the infinitely long memory 'tape' in a Turing Machine. Let $\{\hat{n}_i\}$ be $\hat{m}$ and $\{\hat{m}_i\}$ be $\hat{n}$. Then, the observable $\hat{x}$ denotes the address number of the currently scanned tape location.

The state of $Q$ is a unit vector in the Hilbert space $\mathcal{H}$ spanned by the simultaneous eigenvectors.

$$|x; \hat{m}; \hat{n}\rangle \equiv |n_0, n_1, \cdots, n_{M-1}; \cdots m_{-1}, m_0, m_1, \cdots\rangle$$

Here, the label is an eigenvalue $x$. The state defined in this way is called the *computational basis state*.

We set the spectrum of a two-state observable to be $Z_2$, i.e., the set $\{0, 1\}$ rather than the set $\{-\frac{1}{2}, +\frac{1}{2}\}$ as in physics. The observable with the spectrum $\{0, 1\}$ corresponds to a one-bit element.

The dynamics, i.e., computation of $Q$ can be described by means of a constant unitary operator $U$ on $\mathcal{H}$. Namely, it specifies the evolution of any state $|\psi(t)\rangle \in \mathcal{H}$ at time $t$ in the Schrödinger equation during a single computation step:

$$|\psi(nT)\rangle = U^n |\psi(0)\rangle  (n \in Z^+)$$
$$U^*U = UU^* = 1$$

Here, $U^*$ denotes the unitary conjugate and 1 the unit matrix, respectively.

The computation begins at $t = 0$. At this time, $\hat{x}$ and $\hat{m}$ are 0. The state of finite number of the $\hat{n}$ is a program and input, i.e.,

$$|\psi(0)\rangle = \sum_n \lambda_m |0;0;n\rangle$$

$$\sum_m |\lambda_m|^2 = 1$$

Here, $\lambda_m$ is the $m$-th eigenvalue. Note here that only a finite number of the $\lambda_m$ are non-zero and $\lambda_m$ vanishes whenever an infinite number of the $n$ are non-zero.

To satisfy the requirement that $Q$ operates 'by finite means', the element of the matrix $U$ is of the form:

$$\langle x';m';n' |U|;m;n\rangle = [\delta_{x'}^{x+1} U^+(m',m'_x | m,x_x) + \delta_{x'}^{x-1} U^-(m',m'_x | m,x_x)] \prod_{y \neq x} \delta_{m_y}^{m_y}$$

Here, $\delta_{x'}^{x\pm 1}$ denotes the kronecker symbol. And in each step of a computation, the tape cannot change more than one unit, forwards, backwards, or both.

The functions $U^\pm(m'm' | m,m)$ represent a dynamical motion depending on the local observable $\hat{m}$ and $\hat{m}_x$. And they are arbitrary except that $U$ is unitary. Thus, a choice of $U^\pm$ yields the definition of a different quantum computer $Q[U^+,U^-]$.

Two consecutive states of $Q$ can never be identical. In addition, $Q$ cannot be observed before the computation has ended. A $Q$-program is valid if the expectation value of its running time is finite. Because of unitarity, the dynamics of $Q$, as of any closed quantum systems, are necessarily reversible. On the other hand, the computation of classical Turing Machine is non-reversible. Here, we refer to reversible computation as the computation in which the input can be determined from an output. In fact, classical Turing Machine cannot decide the input from the output mechanically. We will discuss reversible computation in details later.

The universal quantum computer $Q$ has all the properties of the universal Turing Machine. All programs for $Q$ can be expressed by means of the operations on the Turing Machine and the eight operations, which are unitary transformations confined to a single two-dimensional Hilbert space $\mathscr{K}$, i.e., the state space of a single bit.

Let $\alpha$ be any irrational multiple of $\pi$. Then, the ten operations $V_i$ $(0 \leq i \leq 9)$ are as follows.

$$V_0 = \begin{pmatrix} \cos\alpha & \sin\alpha \\ -\sin\alpha & \cos\alpha \end{pmatrix}, V_1 = \begin{pmatrix} \cos\alpha & i\sin\alpha \\ i\sin\alpha & \cos\alpha \end{pmatrix},$$

$$V_2 = \begin{pmatrix} e^{i\alpha} & 0 \\ 0 & 1 \end{pmatrix}, \qquad V_3 = \begin{pmatrix} 1 & 0 \\ 0 & e^{i\alpha} \end{pmatrix},$$

$$V_4 = V_0^{-1}, \qquad V_5 = V_1^{-1},$$

$$V_6 = V_2^{-1}, \qquad V_7 = V_3^{-1},$$

$$V_8 = \frac{1}{\sqrt{2}} \begin{pmatrix} 1 & 1 \\ -1 & 1 \end{pmatrix}, \qquad V_9 = \frac{1}{\sqrt{2}} \begin{pmatrix} 1 & i \\ i & 1 \end{pmatrix}$$

Here, unitary transformation is definable by means of the composition of the operations described above. $V_8$ and $V_9$ are the transformation for $90°$ spin rotations, which is useful to the computation of $Q$.

For each $V_i$, there correspond computational basis elements representing programs $\phi(V_i, a)$, which perform $V_i$ upon the least significant bit of the $a$-th slot.

If $j$ is 0 or 1, these basis elements evolve as follows:

$$| \phi(V_i, 2), j \rangle \rightarrow \sum_{k=0}^{2} \langle k | V_i | j \rangle | \phi(V_i, 2), k \rangle$$

Composition of the $V_i$ may be effected by concatenation of $\phi(V_i, a)$. Thus, there exist programs that effect upon the state of anyone bit a unitary transformation arbitrarily close to any desired one. For each $L$-bit state $| \phi$, there exists a $Q$-program $\rho(| \phi)$ which accurately evolves $| \phi$ to the basis state $| 0_L$ in which all $L$ bits are 0.

The universal quantum computer $Q$ can simulate with arbitrary precision any other quantum computer $Q[U^+, U^-]$, and can also perfectly simulate any Turing Machine. In this sense, it follows that the universal quantum computer is computationally equivalent to the universal Turing Machine. After defining a quantum computer $Q$, Deutsch showed that it can simulate various physical systems beyond the scope of the universal Turing Machine. The examples he presented include random numbers, discrete stochastic systems, quantum correlations, arbitrary physical systems, parallel processing on a serial computer, and faster computers.

Deutsch also discussed further connections between physics and computer science in view of quantum computers. First, he noted that complexity theory for $Q$ is worth investigating. Second, he addressed the connections between CTP and other parts of physics like thermodynamics. Third, he suggested a possibility of programming physics.

Deutsch's proposal of quantum computers seems to give great impact on the area of quantum computing. One of the notable aspects is that he proposed and defended quantum computers by reconsidering CTH to generalize to CTP. Another aspect is that he formally defined a computational model, i.e., $Q$ for quantum computers. Thus, his proposal can be defended conceptually as well as mathematically. However, since his theory was worked out in the position of physicists, it took time until quantum computing is an interesting subject in computer science. We can point out that Deutsch's exposition of the theory of $Q$ is heterogeneous as the presentation of any theory used in computer science and that it appears difficult to understand his theory easily.

### 4.2.3  Formalization of Quantum Turing Machine

To discuss quantum computers in theory of computation, it was necessary to work out a standard theory of quantum computers. By a standard theory, we mean a theory of quantum Turing Machine compatible to classical theory of Turing Machine. In 1993, Bernstein and Vazirani defined a quantum Turing Machine in Bernstein and Vazirani [11].

A quantum Turing Machine is a quantum mechanical generalization of the classical Turing Machine, and can also be interpreted as a version of *non-deterministic probabilistic Turing Machine*. In the quantum Turing Machine, elements like states and symbols are observables in a quantum system. That is, a state is formalized by the vector state in a state space and the evolution (computation) by the unitary operators. And a measurement is performed when the computation stops.

A quantum Turing Machine is, like classical Turing Machine, defined by $QTM = (\Sigma, Q, \delta)$, where $C$ is a set of computable complex numbers.

---

- $Q$ is a finite set of states including the initial $q_0 \in Q$ and the final state $q_f \in Q, q_f \neq q_0 \ (q_0 \neq q_f)$.
- $\Sigma$ is a finite alphabet (finite set of tape with the blank symbol $B \in \Sigma$.
- $\delta : \Sigma \times Q \to C^{\Sigma \times Q \times \{L,R\}}$ is the deterministic quantum transition function, where $L$ denotes the left and $R$ the right.

---

The essential difference of classical Turing Machine and quantum Turing Machine consists in the definition of a transition function. The quantum transition function $\delta(\sigma, q, \tau, q', d) = c$ reads "if $QTM$ reads the symbol $\sigma$ at the state $q$ then it writes the symbol $\tau$ with the *amplitude probability* x, moves the head towards the direction $d$ and changes the state to $q'$. Here, by the amplitude probability we mean a complex number whose modulus squared is a probability.

A computation of $QTM$ is defined in terms of the configuration which consists of the tape symbols, the head position, the state, and the machine position. It can then be described as a tree. Namely, the root of a tree is the initial configuration, the node of a tree is the configuration and the arc is the probability of a transition of a configuration to other configuration. In $QTM$, the path of a tree, i.e., the configuration is simultaneously searched in a computation. The simultaneous search, also called the *quantum parallelism* is done exponentially in computational steps, and is one of the features of quantum computing, yielding ultra high speed computation.

Although several configurations are available in $QTM$ before a measurement, it is possible to have only one configuration when the measurement is performed. The configuration which can be obtained simultaneously corresponds to the superposition state in a machine and it may be an entangled state.

Bernstein and Vazirani proved that the computability of a quantum Turing Machine does not exceed that of probabilistic Turing Machine in Bernstein and Vazirani [11]. For the result, some explanations are in order. As described in Chapter 2, a Turing Machine is usually defined as the so-called *deterministic Turing Machine*. But, a quantum Turing Machine is essentially defined as a non-deterministic Turing Machine as presented above. This implies that the notion of non-determinism is built in the theory in terms of probabilistic concepts. Thus, we must be faced with the so-called *P=NP problem*.

Here, we do not give a detailed treatment of the $P = NP$ problem, which is the problem as to whether polynomial and non-deterministic polynomial computations are the same computational complexity. Therefore, if $P = NP$ holds, it is shown that a quantum Turing Machine is equivalent to a Turing Machine in the computational power. But, if $P \neq NP$, then it is considered that $NP$ belongs to the more complex classes.

This implies that a quantum Turing Machine is equivalent to a probabilistic Turing Machine under certain conditions, and then it can be shown that it is more powerful than the standard Turing Machine. Neglecting the detailed discussions, however, we could roughly say that a quantum Turing Machine is equivalent to a Turing Machine.

A quantum Turing Machine (and Deutsch's $QC$) is a mathematical model for quantum computers, and is used as a tool for quantum complexity theory. Some concepts of complexity depending on quantum computers have been also proposed. But, computational models like a quantum Turing Machine should be regarded as theoreticl models in that they are not always used for the actual algorithm and implementation for quantum computing.

## 4.3   Qubit

We said that the basic unit of information in quantum computing is a *qubit* (quantum bit). It can then be considered as the simplest quantum system. Although Deutsch wrote the concept equivalent to that of a qubit, he did not use the term 'qubit'. As the qubit played an important role in the development of quantum computing, we say more about the qubit below.

### 4.3.1   Bit vs. Qubit

A basic unit of information employed in classical computers is a *bit*. It is either 0 or 1. Namely, if $b$ is a bit, then $b \in \{0, 1\}$. An *n-bit string* consists of the sequence of $n$ bits, i.e., the element of $\{0, 1\}^n$.

Here, the addition of bits is denoted by $\oplus$. Then, $x \oplus y$ is defined as follows.

$$0 \oplus 0 = 1 \oplus 1 = 0$$
$$0 \oplus 1 = 1 \oplus 0 = 1$$

Here, $x + y \pmod 2$ holds. $\oplus$ is also used of the addition of bit strings.

A qubit is a quantum system which can be expressed as the superposition state consisting of two normalized orthogonal states. Two basis states are written as $|0\rangle$ and $|1\rangle$ in the bra-ket notation.

Let $\mathcal{H}$ be the Hilbert space. Then, the general state of qubit, written $|\psi\rangle$, is defined as follows:

$$|\psi\rangle = \alpha |0\rangle + \beta |1\rangle$$
$$\alpha^2 + \beta^2 = 1$$

which denotes the superposition state of $|0\rangle$ and $|1\rangle$. That is, $|\psi\rangle$ simultaneously satisfies $|0\rangle$ and $|1\rangle$.

Note that in 1 qubit, we can interpret it in the following way:

$$|0\rangle = \begin{pmatrix} 1 \\ 0 \end{pmatrix}, \quad |1\rangle = \begin{pmatrix} 0 \\ 1 \end{pmatrix}$$

Consequently, we can see that $|0\rangle$ and $|1\rangle$ in a qubit corresponds to 0 and 1 in a bit, respectively.

The value of a qubit is an observable $\mathcal{N}$ which has the Hamilton operator $N$ on the Hilbert space $\mathcal{H} = C^2$. Here, $N|i\rangle = i|i\rangle$ holds. $N$ can be written in the matrix form as follows:

$$N = \begin{pmatrix} 0 & 0 \\ 0 & 1 \end{pmatrix}$$

The expected value of $N$ is:

$$\langle N \rangle = \langle \psi | N | \psi \rangle = (\alpha^* \quad \beta^*) \begin{pmatrix} 0 & 0 \\ 0 & 1 \end{pmatrix} \begin{pmatrix} \alpha \\ \beta \end{pmatrix} = |\beta|^2$$

where $\alpha^*$ is the unitary conjugate of $\alpha$. By a measurement, $\langle N \rangle$ gives rise to the probability finding a qubit in the state $|\psi\rangle$.

Several qubits are described as the tensor product $\otimes$ of individual qubits. Let the state of a first qubit be $|\phi\rangle$ and the one of a second qubit be $|\psi\rangle$. Then, the state of the composite system of these two qubits becomes $|\phi\rangle \otimes |\psi\rangle$.

The state space of $n$ qubits have $2^n$ states, i.e.,

$$|00\cdots00\rangle, |00\cdots01\rangle, \cdots, |11\cdots11\rangle$$

each of which is called the *computational basis state*. Henceforth, we write $|00\cdots0\rangle$ for $|0\rangle \otimes |0\rangle \otimes \cdots \otimes |0\rangle$.

If $a$ and $b$ are two-dimensional column vectors, then the tensor product $a \otimes b$ is expressed as follows:

$$a \otimes b = \begin{pmatrix} a_1 \\ a_2 \end{pmatrix} \otimes \begin{pmatrix} b_1 \\ b_2 \end{pmatrix} = \begin{pmatrix} a_1 b_1 \\ a_1 b_2 \\ a_2 b_1 \\ a_2 b_2 \end{pmatrix}$$

Thus, if we set:

$$|0\rangle = \begin{pmatrix} 1 \\ 0 \end{pmatrix}, \quad |1\rangle = \begin{pmatrix} 0 \\ 1 \end{pmatrix}$$

then, the following holds:

$$|00\rangle = \begin{pmatrix} 1 \\ 0 \end{pmatrix} \otimes \begin{pmatrix} 1 \\ 0 \end{pmatrix} = \begin{pmatrix} 1 \\ 0 \\ 0 \\ 0 \end{pmatrix}$$

$$|01\rangle = \begin{pmatrix} 1 \\ 0 \end{pmatrix} \otimes \begin{pmatrix} 0 \\ 1 \end{pmatrix} = \begin{pmatrix} 0 \\ 1 \\ 0 \\ 0 \end{pmatrix}$$

$$|10\rangle = \begin{pmatrix} 0 \\ 1 \end{pmatrix} \otimes \begin{pmatrix} 1 \\ 0 \end{pmatrix} = \begin{pmatrix} 0 \\ 0 \\ 1 \\ 0 \end{pmatrix}$$

$$|11\rangle = \begin{pmatrix} 0 \\ 1 \end{pmatrix} \otimes \begin{pmatrix} 0 \\ 1 \end{pmatrix} = \begin{pmatrix} 0 \\ 0 \\ 0 \\ 1 \end{pmatrix}$$

We can write the state of $|\psi\rangle$ of 2 qubits as follows:

$$|\psi\rangle = \alpha|00\rangle + \beta|10\rangle + \gamma|01\rangle + \delta|11\rangle$$
$$|\alpha|^2 + |\beta|^2 + |\gamma|^2 + |\delta|^2 = 1$$

Therefore, $n$ qubits can simultaneously express $2^n$ states. This means that a qubit can express more information than a bit. By this, the fact implies that if we have many qubits then computation speed increases exponentially.

A measurement of $n$ qubits on the computational state is done on its computational basis state, and its result becomes $n$-bit string. If the state is $|\phi\rangle$, then the probability that the result of a measurement on the computational basis is a bit string is given by the square of a scalar product, i.e., $|\langle x|\phi\rangle|^2$. Note, however, that there may be cases that measurement of a state is partial. Thus, the state after a measurement is a state which is compatible to the performed measurement.

Now, we give an example of computation for qubits. Consider the following two qubits state.

$$\frac{1}{\sqrt{3}}|00\rangle + \frac{1}{\sqrt{3}}|01\rangle + \frac{1}{\sqrt{3}}|11\rangle = \frac{1}{\sqrt{3}}|00\rangle + \frac{\sqrt{2}}{\sqrt{3}}\frac{|0\rangle + |1\rangle}{\sqrt{2}} \otimes |1\rangle$$

Here, if we measure the second qubit in the computational basis and the probability that the result of the measurement is 0 is $\frac{1}{3}$, then the state of the first qubit becomes $|0\rangle$ and the probability that the result of the measurement is 1 is $\frac{2}{3}$, then the state of the first qubit becomes $\frac{1}{\sqrt{2}}(|0\rangle + |1\rangle)$.

There are $2^n$ states in $n$ qubits. Therefore, in quantum computing, since quantum parallelism can perform parallel computation, it would be possible to perform ultra high speed computation by increasing the number of qubits as thousands.

## 4.3.2 Polarization of Photons

We can use arbitrary phenomena which have two states for the basis of qubits. Often used are the polarizations of photons and the spin of an electron. Here, we explain the former in some detail. The example of the polarization of photons has been pointed out by Feynman; see Feynman [36] as described before. The state of polarization of one photon can be expressed by means of the unit vector for suitable direction.

Namely, the state $| \psi \rangle$ of arbitrary polarization is given in terms of two base vectors which are orthogonal to each other. For instance, we consider the base vectors whose direction is upper and right, then the state can be written as follows:

$$| \psi \rangle = \alpha | \uparrow \rangle + \beta | \rightarrow \rangle$$

Here, $| \uparrow \rangle$ corresponds to $| 0 \rangle$ and $| \rightarrow \rangle$ to $| 1 \rangle$, respectively. It is here noted that $\alpha$ and $\beta$ are generally complex numbers and that $\alpha^2 + \beta^2 = 1$ holds since the state vector is a unit vector.

Of course, we could consider arbitrary two bases which are orthogonal to each other. Thus, the state of polarization can also be described as follows:

$$| \psi \rangle = \alpha | \nwarrow \rangle + \beta | \nearrow \rangle$$

In quantum mechanics, it is assumed that a measurement device for two-dimensional states has equipped normalized orthogonal base. And a measurement of the state is to transform the state into one of the base vectors of the measurement device. The probability that the state is measured as the base vector $| u \rangle$ is the square of the norm of the amplitude of the original state whose direction is the same as the one of the base vector. Namely, a measurement can be interpreted as the projection on base vectors.

Now, consider the polarization of photons whose base states are $| \uparrow \rangle$ and $| \rightarrow \rangle$. Then, in the measurement of the polarization, the state $| \psi \rangle = \alpha | \uparrow \rangle + \beta | \rightarrow \rangle$ is measured in such a way that $| \uparrow \rangle$ is measured with the probability $| \alpha |^2$ and that $| \rightarrow \rangle$ with the probability $| \beta |^2$. If we employ different measurement devices, the result of a measurement changes because they have naturally different bases. But a measurement is always performed by orthogonal base. It is also to be noticed that if we measure a state then the state changes.

For example, consider the case in which the measurement of $| \psi \rangle = \alpha | \uparrow \rangle + \beta | \rightarrow \rangle$ yields $| \uparrow \rangle$. Then, the second measurement of the same base gives the result $| \uparrow \rangle$ with the probability 1. In other words, we cannot determine the original state from the result of a measurement.

In this way, two-state systems like the polarization of photons can be used for describing qubits. There are other interesting two-state systems for qubits. In addition, the notion of a measurement can be mathematically generalized as the principle of a measurement device in quantum systems. Namely, it can be modeled as the operators in quantum mechanics on the Hilbert space.

## 4.4 Quantum Gates

To construct quantum computers, we need the *circuit* for doing computation. In classical computers, as described in Chapter 2, the circuit is designed by means of the *Boolean algebra*. However, we need a new kind of gates for quantum computers, i.e. quantum gate. The input of a circuit is a bit string $\{0,1\}^n$. *Logic gates* like NOT, OR, AND and NAND transform it to the output. In other words, the output bit can be written as a *Boolean function*, i.e., $f : \{0,1\}^n \rightarrow \{0,1\}$ of the input bit.

In 1988, Deutsch proposed a theory of *quantum gate* in Deutsch [25], whose computation is performed quantum mechanically. It is thus conceptually different from classical logic gate, but it can simulate it. Feynman also developed a similar theory of quantum gates in Feynman [37]. As shown above, the state in quantum computers can be expressed as a qubit. It is labelled by both $|0\rangle$ and $|1\rangle$ as the computational basis. Therefore, qubit corresponds to a bit in the base state.

Actual computation on qubits can be seen as a temporal evolution, and can be interpreted by the unitary transformation on states. This means that computation is specified by the unitary matrix which transforms the state at certain time point into the one at other later time point. The result of computation can be obtained by a measurement defined by the projection on base states. Namely, the base state of the result after the measurement is provided as the probability amplitude which denotes the square of the probability of the result. Consequently, the result of the computation performed by a quantum gate (computer) is a real number.

### 4.4.1  Outline of Quantum Gates

If the above basic procedures of quantum computing are applied to quantum gates, we obtain the following procedures.

---

(1) Initialize qubits as the computational base.

(2) Apply the unitary transformation $U$ to qubits.

(3) Measure the base state and compute the probability amplitude.

---

Thus, the mechanism of quantum gates is displayed as Figure 4.1.

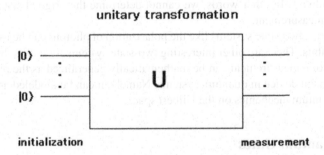

**Fig. 4.1** Mechanism of Quantum Gates

From the figure 4.1, we can understand that quantum gates have to compute the unitary transformation. This can be mathematically formalized in a matrix form.

Let the input qubits be $a$, output qubits be $b$, and the unitary matrix be $U$. Then, a quantum gate is defined by the unitary transformation:

$$b = Ua$$

Since the unitary transformation is reversible, a quantum gate has the same number of inputs and outputs.

The unitary operator $U$ for $n$ cubits is defined on the Hilbert space $\mathscr{H} = C^{2^n}$ as follows:

$$U = \sum_{i=0}^{2^n-1} \sum_{i=0}^{2^n-1} |i\rangle u_{ij} \langle j|$$

where

$$\sum_{k=0}^{2^n-1} u_{ki}^* u_{kj} = \delta_{ij}$$

Thus, quantum gates serve as the device of providing the unitary transformation which is fundamental to quantum computing. There are the following differences of the Boolean function and the unitary operator.

---

- reversibility
- superposition
- parallelism

---

The computation in quantum gates is specified by the unitary matrix $U$, and the computation must be reversible for time since $UU^* = U^*U = E$ holds. Namely, reversible computation is the computation which can reconstruct the input from the output and non-reversible computation cannot. Consequently, in non-reversible computation, information is lost non-reversively. AND and NOT, which are often used as classical logic gates, are non-reversible.

By this means that the computation of traditional computers is not reversible. For example, consider the AND gate which has two inputs and the output 0. Then, we can see that at least one input should be 0. But, we cannot determine which input is 0. Namely, it is impossible to identify the inputs from the output.

### 4.4.2 Useful Quantum Gates

Reversible computation can be interpreted as the computation which contains enough information to determine inputs from outputs. Thus, the computation of quantum computers must be reversible due to the unitarity. For this reason, we have to formalize the classical computation like AND and OR as reversible computation in order to classical logic gates in quantum gates.

Since the 1970's, the problem whether classical computation can be performed reversibly has been independently studied by Bennett and by Fredkin and Toffoli; see Bennett [7, 8] and Fredkin and Toffoli [39]. In the 1970's, Bennett studied the reversibility of computation in Bennett [7]. He showed that the operations on a Turing Machine, i.e., computation can be expressed by reversible computation. To paraphrase the result, it means that computation can be theoretically performed without consuming energy. In the 1980's, he investigated computation in view of thermodynamics; see Bennett [8]. Bennett's basic idea is that computation follows the laws of physics. Namely, we can see that he developed the physics of information. Unfortunately, his theory was not fully appreciated in computer science.

Fredkin and Toffoli investigated practical quantum gates which can compute reversibly, as explained below. Without doubt, Bennett's ad Fredkin and Toffoli's works gave great impact on Feynman's work on quantum mechanical computers; see Feynman [37].

Now, we give a detailed exposition of reversible computation. As a start, we need to address that it is necessary that inputs and outputs are symmetric to describe reversible computation, namely the number of inputs and outputs is equivalent. According to Bennett's and others' work, reversible computation can be defined by the three logic gates, i.e., NOT, CNot and CCNot. This means that there exists the universal gate. Note also that the gates called FANOUT and SWAP (EXCHANGE) are secondarily used.

These gates are depicted by the new notation. As described above, NOT is reversible and we use the notation as in Figure 4.2, since the standard notation in Figure 2.3 is not symmetric.

**Fig. 4.2** NOT Gate

*CNot* (controlled NOT) is described as Figure 4.3, with two inputs and outputs.

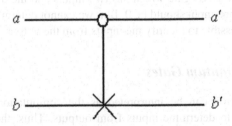

**Fig. 4.3** CNot Gate

Here, the line with ∘ above is a controlled line and the line with × below is an input-output line. If we exchange these lines, the resulting gate is theoretically equal to the original one.

In CNot, unlike ordinary NOT, computation is controlled by the input to the control line. That is, if the input to the controlled line is 0, then the input passes without activating NOT. If the input to the controlled line is 1, then NOT is activated and the input is reversed. Thus, the truth-value table of CNot is as in Table 4.1.

**Table 4.1** Truth-value table of CNot

| $a$ | $b$ | $a'$ | $b'$ |
|---|---|---|---|
| 0 | 0 | 0 | 0 |
| 0 | 1 | 0 | 1 |
| 1 | 0 | 1 | 1 |
| 1 | 1 | 1 | 0 |

From the table, we can see that $b' = a \oplus b$. CNot is reversible in that the identity gate is available if two CNot gates are repeatedly connected.

*CCNot* is called the *controlled-controlled Not*), which also called the *Toffoli gate*. As in Figure 4.4, CCNot has three inputs and outputs.

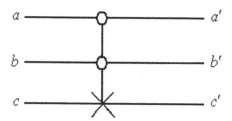

**Fig. 4.4** CCNot Gate

Here, the above two lines are controlled lines and the line below is an input-output line. In CCNot, the inputs in the controlled lines $a$ and $b$ outputs exactly as them and the input-output line $c$ is activated and inputs are reversed if both $a$ and $b$ are 1. Thus, we have the truth-value table of CCNot as in Table 4.1.

From Figure 4.2, we can see that $c' = (a \wedge b) \oplus c$. For example, if we set $c = 0$ then the AND gate can be obtained. Thus, the CCNot gate is the universal reversible gate.

*FANOUT* (fan-out) can branch one input to two outputs, as in Figure 4.5. FANOUT is generally used when we want to increase lines in a circuit.

**Table 4.2** Truth-Value Table of CCNot

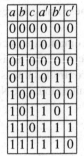

| a | b | c | a' | b' | c' |
|---|---|---|----|----|----|
| 0 | 0 | 0 | 0  | 0  | 0  |
| 0 | 0 | 1 | 0  | 0  | 1  |
| 0 | 1 | 0 | 0  | 0  | 0  |
| 0 | 1 | 1 | 0  | 1  | 1  |
| 1 | 0 | 0 | 1  | 0  | 0  |
| 1 | 0 | 1 | 1  | 0  | 1  |
| 1 | 1 | 0 | 1  | 1  | 1  |
| 1 | 1 | 1 | 1  | 1  | 0  |

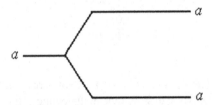

**Fig. 4.5** FANOUT Gate

FANOUT can be represented by means of CNot as in Figure 4.6.

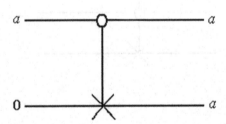

**Fig. 4.6** Representation of FANOUT by CNot

*SWAP*, also called EXCHANGE, exchanges two inputs as in Figure 4.7.

SWAP can expressed by three CNot's as in Figure 4.8. Here, $a' = b$ and $b' = a$ hold.

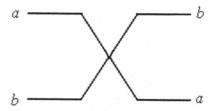

**Fig. 4.7** SWAP Gate

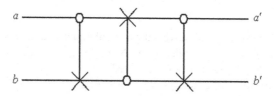

**Fig. 4.8** SWAP by means of CNot

Another universal reversible quantum gate is the *Fredkin gate*, which is a controlled gate as shown in Figure 4.9. It is reversible and has a feature that the number of 0 and 1 in inputs and outputs is invariant.

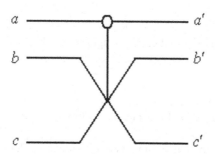

**Fig. 4.9** Fredkin Gate

Here, $a' = a$ holds. In the Fredkin gate, $a$ is a controlled line. If $a = 0$, then $b$ and $c$ are invariant. But, if $a = 1$, then $b$ and $c$ exchange. Thus, the truth-value table of the Fredkin gate is as in Table 4.3.

**Table 4.3** Truth-Value Table of the Fredkin gate

| a | b | c | a' | b' | c' |
|---|---|---|----|----|----|
| 0 | 0 | 0 | 0  | 0  | 0  |
| 0 | 0 | 1 | 0  | 0  | 1  |
| 0 | 1 | 0 | 0  | 0  | 0  |
| 0 | 1 | 1 | 0  | 1  | 1  |
| 1 | 0 | 0 | 1  | 0  | 0  |
| 1 | 0 | 1 | 1  | 1  | 0  |
| 1 | 1 | 0 | 1  | 0  | 1  |
| 1 | 1 | 1 | 1  | 1  | 1  |

The Fredkin gate is also the universal reversible quantum gate. For instance, if $b = 1$ and $c = 0$, then $b' = \bar{a}$, yielding the NOT gate. By setting $c = 0$, we have $c' = a \wedge b$ and the AND gate can be represented.

In this way, we explain the structure of reversible gates. The CCNot gate and the Fredkin gate, which are the universal reversible gates, are often utilized in the study of quantum gates. Deutsch proposed a 3-qubit universal quantum gate, now known as the *Deutsch gate* in Deutsch [26]. The Deutsch gate can simulate classical logical operations.

Now, we explain other quantum gates. Here, we restate the features of quantum gates arising from superpositions. By the unitary transformation $U$, the eigenstate $|\psi\rangle = |k\rangle$ is transformed into the superposition of the eigenstate $|\psi'\rangle$ as follows:

$$|\psi'\rangle = U|k\rangle = \sum_{k'} U_{k'k}|k\rangle$$

If the state $|\psi\rangle$ was already a superposition of several eigenstate, the unitary transformation is simultaneously applied to all eigenstates, that is,

$$U \sum_i c_i |i\rangle = \sum_i c_i U|i\rangle$$

The feature is called the *quantum parallelism*. It is derivable from the linearity of the unitary transformation and the factor of ultra super speed of computation in quantum computing

We now turn to $n$ qubits quantum gates, which can be defined by means of the unitary operators. 1-qubit gates include the following:

---

- *rotation gate*

$$R_\theta = \begin{pmatrix} \cos\theta & -\sin\theta \\ \sin\theta & \cos\theta \end{pmatrix}$$

- *Pauli gate*

$$X = \sigma_x = \begin{pmatrix} 0 & 1 \\ 1 & 0 \end{pmatrix},$$

$$Y = \sigma_y = \begin{pmatrix} 0 & -i \\ i & 0 \end{pmatrix},$$

$$Z = \sigma_z = \begin{pmatrix} 1 & 0 \\ 0 & -1 \end{pmatrix},$$

$$I = \sigma_0 = \begin{pmatrix} 1 & 0 \\ 0 & 1 \end{pmatrix},$$

- *Hadmard gate*

$$H = \frac{1}{\sqrt{2}} \begin{pmatrix} 1 & 1 \\ 1 & -1 \end{pmatrix}$$

- *Phase Shift Gate*

$$R_\theta = \begin{pmatrix} 1 & 0 \\ 0 & e^{i\theta} \end{pmatrix}$$

- $\pi/8$-*gate*

$$T = \sqrt{S} = \begin{pmatrix} 1 & 0 \\ 0 & e^{i\pi/4} \end{pmatrix}$$

---

*Rotation gate* is based on the *rotation*, which is one of linear transformations. $R_\theta$ denotes the rotation around the origin. Deutsch's $V_0$ described before performs the rotation.

*Pauli gate* is the gate utilizes the property of the spin of a particle like an electron. $X, Y$ and $Z$ denote the rotation around the $x, y, z$ axeses, respectively. $I$ denotes the identity transformation. The properties of the Pauli matrices are as explained in Chapter 3. In fact, Feynman discussed the possibility of using the Pauli matrices for quantum computing in Feynman [36].

Note here that Feynman said that the model based on the Pauli matrices can be applied to Bosons but cannot be applied to Fermions but the observation is not correct. In practice, there exist many quantum computers making use of the spin of an electron.

The Pauli matrix $X$ is the operator representing NOT in the following way.

$$X|0\rangle = \begin{pmatrix} 0 & 1 \\ 1 & 0 \end{pmatrix} \begin{pmatrix} 1 \\ 0 \end{pmatrix} = \begin{pmatrix} 0 \\ 1 \end{pmatrix} = |1\rangle$$

$$X|1\rangle = \begin{pmatrix} 0 & 1 \\ 1 & 0 \end{pmatrix} \begin{pmatrix} 0 \\ 1 \end{pmatrix} = \begin{pmatrix} 1 \\ 0 \end{pmatrix} = |0\rangle$$

The Pauli matrix $Z$ is an operator which is regarded as a special case of the phase shift gate as introduced below.

It is noticed that the Pauli matrix $Y$ can be used by simplifying it in the following way:

$$Y = \begin{pmatrix} 0 & -1 \\ 1 & 0 \end{pmatrix}$$

Then, we get the relationships of $X, Y$ and Z.

$$X = YZ,$$
$$Y = ZX,$$
$$Z = XY$$

As said above, the unitary transformation can be applied to superpositions. For example, applying $X$ to $|\psi\rangle = \alpha |0\rangle + \beta |1\rangle$ yields the following:

$$X(\alpha |0\rangle + \beta |1\rangle) = \alpha X(|0\rangle) + \beta X(|1\rangle) = \alpha |1\rangle + \beta |0\rangle$$

Here, the same unitary transformation $X$ is applied to the superposition.

The *Hadmard gate* is the gate which uses the *Hadmard matrix*. Deutsch's $V_8$ is the Hadmard gate. In the Hadmard matrix, each element is either 1 or $-1$, and each row is orthogonal to each other.

In $H$, row elements are normalized. The Hadmard Gate $H$ can generate the superposition states as follows:

$$H|0\rangle = \frac{1}{\sqrt{2}}(|0\rangle + |1\rangle)$$

$$H|1\rangle = \frac{1}{\sqrt{2}}(|0\rangle - |1\rangle)$$

If $H$ is applied to each $n$ bit, then we can generate $2^n$ states, which can be regarded as the binary representation of the numbers from 0 to $2^n - 1$.

$$(H \otimes H \otimes \cdots \otimes H) |00...0\rangle$$

$$= \frac{1}{\sqrt{2}}((|0\rangle + |1\rangle) \otimes (|0\rangle + |1\rangle) \otimes \cdots \otimes (|0\rangle + |1\rangle))$$

$$= \frac{1}{\sqrt{2^n}} \sum_{x=0}^{2^n-1} |x\rangle$$

The quantum gate, which applies $H$ to $n$ bits, is called the *Walsh gate*, denoted $W$. It can be decomposed into the following forms:

$$W_1 = H,$$
$$W_{n+1} = H \otimes W_n$$

The *phase shift gate* uses the difference of the phase of input waveform and output waveform, namely phase shift. Note that $Z$ is defined as the phase shift gate with $\theta = \pi$. The phase shift gate is more understandable if a qubit is defined as the form of wave. Let $|\psi\rangle$ be a general state, $|0\rangle$ and $|1\rangle$ basis states. Then, we can write as follows:

$$|\psi\rangle = \cos\theta |0\rangle + e^{i\phi} \sin\theta |1\rangle$$

where $\phi$ denote the *phase* between $|0\rangle$ and $|1\rangle$. In the definition, we assume *Euler's formula* concerning the relation of complex numbers and trigonometric functions, as described below:

$$e^{i\theta} = \cos\theta + i\sin\theta$$

If we measure $|\psi\rangle$, then the result is that the probability of $|1\rangle$ is $\sin^2\theta$ and the probability of $|0\rangle$ is $\cos^2\theta$. The phase-shift gate is assumed to be of use for quantum communication. $\pi/8$-*gate* is a special case of the phase shift gate, with $\theta = \frac{\pi}{4}$.

There are the following 2-qubits gates:

- *CNot Gate*
  $CNot : |x,y\rangle \rightarrow |x\oplus y,y\rangle$
- *SWAP Gate*
  $SWAP : |x,y\rangle \rightarrow |y,x\rangle$
- *Controlled Phase Gate*
  $CPhase : |x,y\rangle \rightarrow i^{xy}|x,y\rangle$

The following are the examples of 3-qubits gates.

- *Toffoli gate* (CCNot gate)
  $CCNot : |x,y,z\rangle \rightarrow |x\oplus(y\wedge z),y,z\rangle$
- *Fredkin gate*

$$CSwap : |x,y,z\rangle \rightarrow \begin{cases} |y,x,z\rangle & (z=1) \\ |x,y,z\rangle & (z=0) \end{cases}$$

- *Deutsch gate*

$$D(\theta) : |x,y,z\rangle \rightarrow \begin{cases} i\cos\theta\,|x,y,z\rangle + \sin\theta\,|x,y,1-z\rangle & (x=y=1) \\ |x,y,z\rangle & (\textit{otherwise}) \end{cases}$$

Some of these gates have been already discussed. The Toffoli gate can be interpreted as the Deutsch gate $D(\frac{\pi}{2})$. As a consequence, the universal quantum gate can simulate the universal gate.

One of the important 2-qubits gates is CNot, which can be defined by means of the $X$ gate as follows:

$$CNot = \begin{pmatrix} E & 0 \\ 0 & X \end{pmatrix} = \begin{pmatrix} 1 & 0 & 0 & 0 \\ 0 & 1 & 0 & 0 \\ 0 & 0 & 0 & 1 \\ 0 & 0 & 1 & 0 \end{pmatrix}$$

Also noted is that the Toffoli gate can simulate arbitrary classical logical gates.

From the above exposition, quantum gates can perform various computations. But, from the properties in quantum mechanics, we cannot generate the clone of qubits. In other words, we have no quantum gates capable of copying qubits.

### 4.4.3  No-Cloning Theorem

The property that qubits cannot be copied is proved as the so-called *no-cloning theorem* see Wootters and Zurek [86]. At a glance, it seems inconvenient, but it plays an important role in quantum codes.

The no-cloning theorem can be derived as a consequence of the property of the unitary transformation. Now, assume that we copy a state $|\psi\rangle_A$ in a quantum system $A$. For this purpose, we need another quantum system $B$ which has the state space same as the one in $A$ and has a basis state $|e\rangle_B$ which is independent to $|\psi\rangle_A$. Then, the composite system of $A$ and $B$, denoted $A \otimes B$, has the following states:

$$|\psi\rangle_A |e\rangle_B$$

Here, let the unitary operator generating the clone be $U$. Then,

$$U(|\psi\rangle_A |e\rangle_B) = |\psi\rangle_A |\psi\rangle_B$$
$$U(|\phi\rangle_A |e\rangle_B) = |\phi\rangle_A |\phi\rangle_B$$

hold for all $|\psi\rangle$ and $|\phi\rangle$. Since the unitary operator is closed under the scalar product *, we can have that:

$$\langle e|_B \langle \phi|_A U^*U |\psi\rangle_A |e\rangle_B = \langle \phi|_B \langle \phi|_A |\psi\rangle_A |\psi\rangle_B$$
$$\langle \phi|\psi\rangle = \langle \phi|\psi\rangle^2$$

which implies that $\psi = \phi$ or $\psi$ and $\phi$ are orthogonal. But they do not hold in general. Therefore, it can be shown that there are no unitary operators generating the clone of qubits.

For example, let two orthogonal states be $|a\rangle$ and $|b\rangle$ and the unitary operator for cloning be $U$. Then, the following hold.

$$U(|a0\rangle) = |aa\rangle$$
$$U(|b0\rangle) = |bb\rangle$$

Next, we set:

$$|c\rangle = \frac{1}{\sqrt{2}}(|a\rangle + |b\rangle)$$

and apply $U$ to it. Due to the linearity of the unitary transformation, we obtain:

$$U(|c0\rangle) = \frac{1}{\sqrt{2}}(U(|a0\rangle) + U(|b0\rangle))$$
$$= \frac{1}{\sqrt{2}}(|aa\rangle + |bb\rangle)$$

But, if we compute $U(|c0\rangle)$ by definition, then

$$U(|c0\rangle) = |cc\rangle = \frac{1}{2}(|aa\rangle + |ab\rangle + |ba\rangle + |bb\rangle))$$

can be obtained. However, it is different from the above $U(|c0\rangle)$. Therefore, we can see that qubits cannot be copied. The reason that we cannot copy qubits is that the states in qubits change by a measurement. Actually, in quantum mechanics, a measurement is performed probabilistically as well as destructively.

Finally, we discuss the *universal quantum gate*. As you know, every Boolean function can be computed by the universal gate based on AND and NOT or on NAND. Similar results have been obtained for quantum gates. As described above, in 1989, Deutsch proposed 3-qubits universal quantum gate, i.e., Deutsch gate; see Deutsch [25]. It has also been shown that there is a universal quantum gate for 2-qubits in Deutsch, Barenco and Ekert [25]. In fact, any unitary transformation can be decomposed by the sequence of gates consisting of 1-qubit gates and CNot's. However, the universal quantum gate based on $\{T, H, CNot\}$ can be also utilized for convenience.

## 4.5 Shor's Algorithm

As explained above, the computation in quantum computers is quantum mechanically performed. In practice, quantum computers have some advantages over classical computers in various respects. However, to solve concrete problems by a quantum computer, we need *quantum algorithms*, which can be described as sequences of state transformations and a measurement after the transformations.

In 1994, Shor proposed a quantum algorithm for prime factorization in Shor [72, 73]. His algorithm was a breakthrough, and it is now called the *Shor's algorithm*. We can say that to appreciate the importance of quantum computing is due to Shor's algorithm.

So far, it can be recognized that there exist no algorithms to solve prime factorization in high speed by classical computers. The property constituted the basis of the safety of the *RSA code*, as proposed by Rivest, Shamir and Adelman [67] in 1978. However, the situation changed by Shor's algorithm. In fact, it will be possible to perform high speed computation to solve prime factorization by means of Shor's algorithm. By a high speed algorithm, it generally means the algorithm which can be computed in polynomial time.

By Shor's algorithm, there has been a possibility for decoding the RSA code. In other words, the safety of the RSA code, which has been considered safe, will be problematic, if quantum computers are implemented in practice. The news was shocking, but it stimulated interest in quantum computing and activated its research. In fact, Shor says in Shor [73]:

> it is tempting to speculate that integer multiplication itself might be speeded up by a quantum algorithm; if possible, this would result in a somewhat faster asymptotic bound for factoring on a quantum computer, and indeed could even make breaking RSA on a quantum computer asymptotically faster than encrypting with RSA on a classical computer.

### 4.5.1 RSA Code

Before giving an exposition of Shor's algorithm, we simply rehearse the RSA code. The reader is referred to Rivest, Shamir and Adelman [67] for detailed presentation of the RSA code.

Now, we outline the RSA code. First, we need to select two big integers, say, $p$ and $a$, and compute their composite number $n$, i.e.,

$$n = pq$$

where $p$ and $q$ are kept in secret.

Next, compute the Euler function $\varphi(n) = (p-1)(q-1)$, and find the pair $(e, d)$ satisfying:

$$\gcd(d, \varphi(n)) = 1,$$
$$ed \equiv 1 \bmod \varphi(n)$$

Namely, $d$ is an integer coprime with $\varphi(n)$ and $e$ is the reciprocal of $d$. $d$ can be computed by the extended Euclid algorithm. Here, the public key is the pair $(n, e)$ and the secret key is $d$. $n$ is called the *RSA module*, $e$ is called the *encryption exponent*, $d$ is called the *decryption exponent*, respectively.

*Encryption* and *decryption* are formalized as follows. In encryption, a sender $A$ describes a plain text $m$ by an integer such that $0 \leq m < n$, and by the receiver $B$'s public key $(n, e)$, set the cipher text $c$ as:

$$c = m^e \pmod{n}$$

where $c$ is the integer such that $0 \leq c < n$. Thus, encryption function $E_e$ is defined as follows:

$$E_e = m^e \pmod{n}$$

where $e$ is $B$'s public key.

In decryption, $B$ uses the secrete key $d$ which only $B$ knows and obtain $m$ from:

$$m = c^d \pmod{n}$$

Thus, decryption function $D_d$ is defined as follows:

$$D_d = c^d \pmod{n}$$

This completes the description of the RSA algorithm.

### 4.5.2 Description of Shor's Algorithm

Next, we describe Shor's algorithm shortly. Note that to understand Shor's algorithm, we need some background of number theory and Fourier transform. In a computer, the multiplication of integers is easy to compute, but its inverse operation, i.e., *factoring* is not. It is thus impossible to perform factoring of big integer $N$ fastly.

Shor's idea is to replace the problem of factorization by the problem of *period finding* and solve it by means of the *quantum Fourier transform*. Shor's study was influenced by Simon's algorithm finding the periodicity of a function, which was proposed around the same time; see Simon [75]. Shor used the method of factoring making use of the property that remainders have certain periodicity, which has never been applied to the previous algorithms for factorization.

Let $a$ and $b$ be integers. We denote its greatest common divisor (GCD) by $\gcd(a,b)$. Then, his algorithm for finding the factorization of integer $N$ can be written as follows:

---

**Shor's Algorithm for Factorization**

(1) Select a suitable integer $x$ such that $x < N$.

(2) Compute $f = \gcd(x,N)$. If $f \neq 1$, then $f$ is a factor of $N$. Otherwise go to (3).

(3) Find the least integer $r$ such that its remainder is 1 when dividing $x^f$ by $N$, i.e., $x^f \bmod N = 1$.

(4) If either $\gcd(x^{\frac{r}{2}} - 1, N)$ or $\gcd(x^{\frac{r}{2}} + 1, N)$ is not 1, then it is the factor. Otherwise go to (1).

---

Here, in (3), $r$ is called the *order*. $\gcd(a,b)$ can be computed by the *Euclid algorithm* in polynomial time.

Now, perform factoring of 15 by the above algorithm. In 2001, IBM succeeded in factoring 15 using 7 qubits quantum computer. First, by (1), select a suitable integer $11 < 15$.

By (2), find $\gcd(11,15)$. Then, we have $f = 1$ and go to (3). By (3), find $r$. Since we have:

$$r = 1 : 11^1/15 = 0...11$$
$$r = 2 : 11^2/15 = 8...1$$

the solution is $r = 2$.

By using (4), we have:

$$x^{\frac{r}{2}} - 1 \ 11^1 - 1 = 10$$
$$x^{\frac{r}{2}} + 1 \ 11^1 + 1 = 12$$

Then, compute $\gcd(10,15)$ and $\gcd(12,15)$. As we obtain $\gcd(10,15) = 5$ and $\gcd(12,15) = 3$, the desired factors are 3 and 5. Therefore, the factorization of 15 is: $15 = 3 \times 5$.

### 4.5.3 Discrete Fourier Transform and Quantum Algorithm

We here focus on an interesting feature of Shor's algorithm. This is the procedure (3). In Shor's algorithm. it is possible to fastly find $r$ by means of the quantum algorithm for discrete Fourier transform.

For certain integer $q$, consider the integer $a$ satisfying $0 \leq a < q$. Then, the discrete Fourier transform $A_q$ can be defined as follows:

$$A_q : |a\rangle \rightarrow \frac{1}{\sqrt{q}} \sum_{c=0}^{q-1} |c\rangle \exp(2\pi iac/q)$$

Here, $A_q$ denotes the unitary matrix whose $(a,c)$-element is $\frac{1}{\sqrt{q}}\exp(2\pi ac/q)$. Shor set $q = 2^l$ as $A_q$ can be computed in polynomial time. By this, we can describe the integer $a$ as $|a_{l-1}a_{l-2}\cdots a_0\rangle$.

We look at how the order is computed in Shor's algorithm. First, for integers $n$ and $x$, find the order of $q$ $(n^2 \leq q < 2n^2)$. Since these values are invariant, they can be built in quantum gates.

Next, store the integer $a \pmod q$ in register1 as a superposition. Here, the state is as follows:

$$\frac{1}{\sqrt{q}} \sum_{a=0}^{q-1} |a,0\rangle$$

Next, compute $x^a \pmod n$ by register2. Because $a$ is in register1, the computation is done reversively. Then, the resulting state becomes:

$$\frac{1}{\sqrt{q}} \sum_{a=0}^{q-1} |a, x^a \pmod n\rangle$$

Next, apply the Fourier transform $A_q$ to register1, i.e.,

$$|a\rangle \rightarrow \frac{1}{\sqrt{q}} \sum_{c=0}^{q-1} \exp(2\pi iac/q) |c\rangle$$

Then, the state is:

$$\frac{1}{\sqrt{q}} \sum_{a=0}^{q-1} \sum_{c=0}^{q-1} \exp(2\pi iac/q) |c, x^a \pmod n\rangle$$

Finally, measure the state of register1. If we assume that $0 \leq k < r$, the probability that $|c, x^k \pmod n\rangle$ is:

$$\left| \frac{1}{q} \sum_{a:x^a \equiv x^k} \exp(2\pi iac/q) \right|^2$$

Hence, from the probability, we can find $r = x^k \pmod n$.

Since (4) in the above algorithm can be computed classically, combining these computations enables factorization. Note that Shor's algorithm is not an algorithm which can be defined as a black box. But the computation of the order, whose fast computation has been regarded impossible, is done in polynomial time.

In fact, Shor's algorithm is an efficient (fast) algorithm in which the prime $N$ can be factored in the order $\log N$. This is in fact a breakthrough. But, Shor evaluated his work in the end of Shor [73]:

> Finding polynomial-time algorithms for solving these problems on a quantum computer would be a momentous discovery. There are some weak indications that quantum computers are not powerful enough to solve NP-complete problems [Bennett et al. 1994], but I do not believe that this potentiality should be ruled out as yet.

where "these problems" refers to the NP-complete problems. By now, any quantum computer did not contribute to the attempts to solving the $P = NP$ problem. But we could say that the use of a quantum computer is very attractive.

## 4.6 Other Quantum Algorithms

Indeed Shor's algorithm was shocking, and various useful quantum algorithms have been also proposed. Some of such quantum algorithm can be listed as follows:

---

- Grover's algorithm
- HS algorithm
- Quantum walk
- Adiabatic quantum algorithm

---

Below we will briefly deal with these algorithms.

### 4.6.1 Grover's Algorithm

*Grover's algorithm* is an algorithm for *search* of data in unstructured *database*; see Grover [44]. In unstructured database, data are not sorted structurally. Naturally, the search for unstructured database needs more computational cost than in structured database. The algorithm for fast search for data in database is an important problem.

It is known that the complexity of the classical algorithm, i.e., linear search for data in the unstructured database with $n$ data is $O(n)$. In contrast, if we employ Grover's algorithm, the complexity can be reduced to $O(\sqrt{n})$. If $n$ becomes bigger, the difference to classical algorithms will be clearer.

Because quantum computing supposes the case in which huge number of data should be processed, Grover's algorithm is more useful than classical algorithm for such cases. Thus, Grover's algorithm is, like Shor's algorithm, to address the advantages of quantum algorithms.

The problem which Grover's algorithm can handle is presented as follow. First we define a function $f$: $f : A \rightarrow \{0, 1\}$, where $A$ is a set with $N = 2^n$ elements. There is only an element $s \in A$ with $f(s) = 1$. The problem is to look for the element $x$ such that $f(x) = 1$.

The quantum operator, also called *oracle*, is defined as follows:

$$|x\rangle \rightarrow |x\rangle \text{ (first register)}$$
$$|y\rangle \rightarrow |y \otimes f(x)\rangle \text{ (second register)}$$

where the second register is initialized with $\frac{1}{\sqrt{2}}(|0\rangle - |1\rangle)$.

Then, we can write Grover's algorithm as follows:

---

**Grover's Algorithm**

(1) Initialize the first register in the superposition state:
$\left(\frac{1}{\sqrt{N}}, \frac{1}{\sqrt{N}}, \ldots, \frac{1}{\sqrt{N}}\right)$

(2) Repeat the following operation $O(\sqrt{N})$ times:

(a) Apply the quantum operator.

(b) Apply the matrix $D_{ij} = -\delta_{ij} + 2\frac{1}{N}$

(3) Measure the resulting state of the fist register.

---

For the details of Grover's algorithm, the reader is referred to Grover [44].

### 4.6.2   HS Algorithm

*HS algorithm* (Hidden Subgroup algorithm) is a generalization of Shor's algorithm, which determines the subgroup in a group. The *HS problem* is one of the most important problems in mathematics and computer science, and is closely related to such problems as the prime factorization and the graph isomorphism problem; see Jozsa [52].

The function $f$ in the period finding problem over $Z_M$ is constant on sets $\{x, x + a, \ldots\}$ for each $x$ and distinct on disjoint such sets. If $a$ divides $M$, then it is constant on cosets $x + \langle a \rangle$ of the subgroup of $Z_M$ generated by $a$ and distinct on different such cosets. The HS problem is defined as follows. Given a function $f : G \rightarrow R$ on a group $G$, and a subgroup $H < G$ such that $f$ is constant on (left) cosets of $H$ and distinct for different cosets, and a set of generators for $H$.

The HSP is an important problem. An efficient algorithm for the group $Z_M$ subsumes an efficient factorization algorithm. In HS algorithm, quantum Fourier transform is generally applied. But it is still open whether there exists an efficient HS algorithm without using quantum Fourier Transform.

### 4.6.3   Quantum Walks

*Quantum walk* is a quantum algorithm for *random walk*, which is the movement the next position is determined probabilistically. For example, the so-called *Brown movement* is a typical example. Random walk is also applied as a method for *simulation*. In classical algorithms for random walk, the randomness and probabilistic algorithms are generally used. Depending on the structure of time, there are two kinds of random walks, namely *discrete random walk* and *continuous random walk*.

Quantum walks can improve classical algorithms; see Kempe [54]. In quantum walks, the position of a walker is represented as a superposition. A continuous quantum walk algorithm was proposed by Farhi and Gutman in 1998; see Farhi and Gutman [35]. In their algorithm, the unitary transformation is directly applied to the space in which random walks are performed.

*Ambainis' algorithm* also handles discrete quantum walks; see Ambainis [2]. In discrete quantum algorithm, we need a register for quantum flip-flop. Quantum walks gives rise to the best algorithm for solving the 3SAT algorithm in Boolean algebras. Also noted is that by using quantum walks, we can perform search faster in Grover's algorithm.

### 4.6.4 Adiabatic Quantum Algorithm

The *adiabatic quantum algorithm* was introduced by Farhi and others as a quantum algorithm for optimization; see Farhi et al. [34]. The algorithm can be interpreted as another formalization of the Schrödinger equation. In adiabatic quantum algorithm, the problems of optimization and constraint satisfaction are encoded as local constraints, i.e., the sum of Hamilton operators $H = \sum_i H_i$. The basis state of $H$ does not satisfy many constraints, and represents an optimal solution.

To obtain such a state, different Hamilton operator $H'$ is chosen to easily describe a basis state $| \psi' \rangle$. And the Hamilton operator slowly changes from $H'$ to $H$. By the adiabatic theorem, the state $H(t)$ at time $t$ becomes a basis state, and we can obtain a solution.

Quantum mechanically, the adiabatic quantum algorithm firstly sets a Hamiltonian and makes continuously change the state of a system by means of the Schrödiner equation. And the state finally obtained, i.e., wave function expresses a solution of the problem of optimization. Thus, in the adiabatic quantum algorithm, we need to define a Hamiltonian in which an initial state depends on time so that the final state represents a solution.

It is known that the adiabatic algorithms have computational efficiency in the same way as Grover's algorithm. It is also possible to simulate arbitrary quantum gates in terms of a suitable adiabatic algorithm; see Aharonov [1]. Consequently, the adiabatic algorithms and quantum gates can be considered as equivalent quantum models. In fact, the adiabatic algorithms are founded on the Schrödinger equation and standard quantum algorithms like quantum gates on the Heisenberg equation.

# Chapter 5
# Applications of Quantum Computing

**Abstract.** Chapter 5 outlines the representative applications of quantum computing. We take up quantum codes, quantum communications, quantum teleportation and quantum programming.

## 5.1 Quantum Codes

This chapter is concerned with some applications of quantum computing. Since these applications are now under way, we explain main approaches. If a quantum computer is realized and super high speed computation is possible, then the safety of current coding systems like the RSA code will not be guaranteed. In fact, Shor's algorithm showed such a situation would come true. For example, the securty of *internet* will be problematic. Consequently, we need a new type of code which cannot be decoded by any high speed computers. Quantum computing could solve the difficulty.

*Quantum code* is seen as a code for the next generations to accommodate to the situation. Quantum code is not to guarantee the safety by huge computational cost but to do it by verifying whether the encryption key of a code is tapped on the basis of Heisenberg's uncertainty principle. As we explained in Chapter 2, two physical values cannot be precisely determined simultaneously according to the uncertainty principle. Thus, if we try to measure one physical value precisely, we cannot precisely measure the other physical value. This is the starting point of quantum code.

If we make use of the uncertainty principle, wire tapping of other people can be recognized and the state changes. Therefore, it is impossible to copy the key. Quantum code is one of the typical examples of applications of quantum communications as described below.

### 5.1.1 Quantum Key Distribution

A basic idea of quantum code can be found in Wiesner's *conjugate coding*, which was proposed in the late 1960's; see Wiesner [84]. Later, in 1984, Bennett and others

S. Akama, *Elements of Quantum Computing*,
DOI: 10.1007/978-3-319-08284-4_5, © Springer International Publishing Switzerland 2015

proposed a theory of *quantum key distribution*; see Bennett, Brassard and Ekert [9]. Now, the quantum key distribution is called the *BB84 protocol*. In the quantum key distribution, single photon system is adopted and two orthogonal states of polarization are assigned to bit values. Formal sender and receiver randomly use conjugate polarization basis. Then, it constructs a protocol in which a third person cannot access to information without remaining evidence for the access.

As Shor's algorithm was proposed in 1994, the safety of the open key public code systems like the RSA code cannot be guaranteed. As a consequence, the research of quantum codes became active. Now, it is regarded that the quantum key distribution is one of the effective methods for quantum codes. It is also noticed that since the late 1990's the implementation of system for quantum codes using the devices like optical fiber was investigated.

## 5.1.2   Example of Quantum Code

Here, we give an example of quantum codes. Let $A$ be a sender, $B$ be a receiver and $C$ be a tapper. We assume that information is transmitted in an optical communication channel. $A$ sends $B$ the common key in encrypted form, and the communication can be interpreted as safe if the common key cannot be tapped. This feature is the important point in quantum codes.

A common key is a random bit sequence, and $A$ encrypts a message by it. $B$ decrypts the message by the same common key. Here, the common key is sent from $A$ to $B$ by means of a safe method. Even if the message from $A$ to $B$ is tapped, $C$ cannot decode it unless he knows the common key. For a safe sending of the common key, quantum codes are utilized.

Previous public key codes are based on the difficulty of prime factorization, but the point in quantum codes is to make use of the uncertainty principle in quantum mechanics. A quantum code system generally consists of the classical channel and the quantum channel.

The classical channel sends encrypted messages and information generating a common key. There is no problem if $C$ taps the information in the classical channel. In the quantum channel, quantum code is sent. For instance, if we assume that a quantum code system based on single photon then the optical communication channel is used. Here, for the direction of the movement of single photon we can specify four polarization states, i.e., horizontal direction, vertical direction, $45°$ direction, and $-45°$ direction as quantum states.

Such quantum states based on single photon can be found in Feynman [36]. In BB84, $A$ assigns 1 and 0 to these four quantum states, and sends the message described by random pulse wave by 0 and 1 as the sequence of single photon in polarization states. $B$ reads the single photon sequence by randomly using the device which can recognize horizontal and vertical polarization states and the device which can recognize $\pm45°$ polarization states. In addition, $A$ sends $B$ the information whether horizontal and vertical polarization states or $\pm45°$ polarization states are used, a bit sequence with the signal in which the information agrees is adopted as a common key. Finally, $B$ decrypted the message by means of the common key which is generated in this way.

There are several methods which $C$ taps a quantum code. One method is to directly tap photons. But, quantum states changes in this case, and the tapping can be detected by the uncertainty principle. In the case in which single photon is used, this is to tap a photon and its detection is easy. Another method is $C$ detects the polarization state by the device same as the one $B$ uses and send $B$ the photon in the same polarization state. But, $C$ cannot know which polarization states are use in the sending a message. Thus, the probability $C$ can correctly tap the message is 50%.

In addition, we can protect this type of tapping by making use of the technique of error detection and correction in coding theory. Namely, we add detection bits for tapping to the message from $A$ to $B$. If the probability of error detection is high, we judge that a tapping is done and discard the common key. Further strengthening of quantum codes is available by using the polarization states of two-photon system. This makes the communication channel long. It also has the advantage that entangled states are possible.

For two photons in entangled states, if one photon is in the vertical polarization state then the other photon is in the horizontal polarization state. But, it is impossible to determine the polarization state before a measurement. Unless a measurement is performed, the entangled state continues even if two photons are separated and the tapping becomes more difficult. Consequently, it is considered that the quantum code based on the entanglement of two photons is more promising.

## 5.2 Quantum Communications

*Quantum communication* is information communication which uses the idea of quantum computing. It is regarded as one of the important topics in *quantum information theory*.

Quantum communications are very attractive in that high speed and capacity of information communication and error correction are possible. Observe also that quantum teleportation is classified as the topic in quantum communications in a broad sense.

In quantum communication, it is important to send classical information as quantum states. From Shannon's theorem, maximum data transfer speed is determined for classical communication, but quantum communications have a possibility of transcending its limitation.

**Claude Shannon (USA: 1916-2001)**

Claude Shannon was born in Petoskey, Michigan, USA on 1916. He is referred to as the father of information theory. In 1932, he entered the University of Michigan. He received M.Sc. degree in 1937 from MIT for the study of the use of Boolean algebra for electrical circuit designs. He received Ph.D. degree from MIT in 1940. In the same year, he became a National Research Fellow at the Institute for Advanced Study at Princeton. In 1948, he published information theory in which the concept of information was first defined. He also contributed to the areas like cryptography and computer chess. Shannon died in Medford, Massachusetts, USA on 2001.

It is common knowledge that current information communication technology is based on Shannon's *information communication model*; see Shannon [68]. In his information communication model, information is exchanged as a sequence of symbols, i.e. a bit string. On the other hand, in quantum communications, information is described as quantum states. Thus, in a quantum communication channel, communication channel is described as a mapping from the quantum state for an input to the quantum state for the output. Then, coding is regarded as the transformation of the alphabets to quantum states. And decoding is to find the correspondence of the result of measuring quantum states to alphabets.

## 5.2.1 Holevo Bound

The problem whether the limit of *channel capacity* can be extended was studied in the 1960's by workers like Holevo; see [49]. Holevo formalized the notion of the so-called *Holevo bound*. In 1998, Holevo proved that the Holevo bound and channel capacity agree in connection with quantum information communications; see Holevo [50]. In fact, Holevo showed that at most *n* bits of classical information can be carried by a quantum system of *n* distinguishable qubits, i.e., two-dimensional quantum systems.

But, the Holevo bound includes arbitrary encoding and decoding strategies. Consequently, for information transmission with infinite dimensional systems through noiseless channels, the Holevo bound predicts infinite capacities. The Holevo bound is an upper bound on the amount of classical information that can be accessed from a quantum ensemble in which the information is encoded.

Assume that a sender $A$ obtains the classical message $i$ and he knows that this happens with the probability $p_i$. He wants to send it to a receiver $B$. To do so, $A$ must encode the information $i$ in a quantum state $\rho_i$ and sends the quantum state to $B$. $B$ receives the ensemble $\{p_i, \rho_i\}$ and wants to obtain as much information as possible about $i$, giving the result $m$ with the probability $q_m$ by a measurement. The classical information $I(i:m)$ thus obtained is expressed as follows:

$$I(i:m) = H(p_i) - \sum_m q_m H(p_{i|m})$$

where $H(r_x) = -\sum_r x \log_2 r_x$ denotes the Shannon entropy of the probability distribution $r_x$.

The receiver $B$ wants to obtain the maximal information. It is called the *accessible information*, denoted by $I_{acc}$.

$$I_{acc} = \max I(i:m)$$

where the maximization is over all measurement strategies. However, it is difficult to compute $I_{acc}$.

Holevo introduced the universal bound called the Holevo bound denoted $\chi(p_i, \rho_i)$ and showed the relation:

$$I_{acc}(p_i, \rho_i) \leq \chi(p_i, \rho_i) = S(\overline{\rho}) - \sum_i p_i S(\rho_i)$$

where $\overline{\rho} = p_i \rho_i$ is the average ensemble state and $S(\rho) = -\mathrm{tr}(\rho \log_2 \rho)$, known as the von Neumann entropy.

Now, consider a quantum channel $\mathscr{R}$ which acts on $d$-dimensional quantum systems as inputs. Assume that $A$ wants to send classical information $i$ that happens with the probability $p_i$ by the quantum channel to $B$. He encodes this classical information in the quantum state $\rho_i$ whose Hilbert space is $d$-dimensional.

The classical capacity of this quantum channel is the maximum classical information which can be sent by this channel, and is therefore the accessible information of the ensemble $\{\rho, \mathscr{R}(\rho_i)\}$.

$$C = \max I_{acc}(\{\rho_i, \mathscr{R}(\rho_i)\})$$

where $C$ is maximized over all specified ensembles on the $d$-dimensional Hilbert space $\mathscr{H}^d$. By the Holevo bound, we have:

$$C \leq \max; \left( S(\mathscr{R}(\overline{\rho})) - \sum p_i S(\mathscr{R}(\rho_i)) \right)$$

We can also consider the case of sending qubits through quantum channels. However, we do not go into details here. By the late 1990's, foundations for quantum communication theory have been established by Shor. Recently, a qubit used in quantum communications is called the *flying qubit*, and is distinguished from the *stable qubit* used in local quantum computers. For example, a photon is a flying qubit.

One of the important techniques in information communication is an error correction. It has been considered over many years that *quantum error correction* is difficult. This is because if we try to measure the state of qubits then the state changes. Therefore, the concept of error correction is meaningless. However, Shor and Steane showed that quantum error correction is possible; see Shor [74] and Steane [76].

### 5.2.2 Quantum Teleportation

*Quantum teleportation* is a method of transmitting qubits from one place to another. This is done by combining entangled states and classical communication channels. In 1993, Bennett showed that quantum teleportation is theoretically possible; see Bennett et al. [10]. Quantum teleportation can be interpreted as an application of the *EPR paradox* which was explained in Chapter 3. If two particles are generated, they reach an entangled state in which these two particles are related each other. The pair of these particles in such a state is called the *EPR pair*.

If a measurement of the state of one particle in the EPR pair is performed, then the state of the other can be immediately determined. However, to determine the state of two far separated particles is inconsistent the principle of light speed, and Einstein and others considered such phenomena as a paradox. Bennett and others proposed a theory of quantum teleportation, but it took more than ten years for its experimental justification. In fact, other people have shown teleportation experimentally in various systems including photons, coherent light fields and nuclear spins.

We here stress that in quantum teleportation particles are transferred spatially. By quantum teleportation, a state of certain particle can determine the states of other separated particles. However, quantum teleportation can be applied to quantum communications. For a receiver at separated place can reproduce the unknown state of a sender.

For instance, consider that a sender $A$ generates two photons which forms the EPR pair. Next, he sends a receiver $B$ the photon-1 by the quantum channel. Next, $A$ measure the photon-1. Then, the state of photon-1 is determined and therefore the state of photon-2 is also determined. Consequently, $B$ can know the state of photon-2 by sending $B$ the state of photon-2 through classical channel. Here, although $A$ sends $B$ the unknown state, this state changes after a measurement.

Thus, the unknown state is not copied. Also, the use of classical channel preserves causality. From the discussion, $B$ can reproduce the state which $A$ sent to $B$. At a glance, quantum teleportations appear science fictional. But they are the technology on the basis of the principle of quantum mechanics described above. For the future, quantum teleportations will play a crucial role in the study of quantum communications.

The entanglement of photons and the angular momentum are mainly used as the bases for theories of quantum teleportation. However, there are other methods for quantum teleportation. Recently, quantum teleportations of continuous values are also investigated.

In 1997, the research group of Innsbruck University first succeeded in the experiment of quantum teleportation; see Bouwmeester et al. [17]. Since then, various types of quantum teleportation have been experimentally justified. Recently, Furusawa and others at University of Tokyo succeeded in the experimental teleportation of strongly non-classical wave packets of light; see Lee et al. [58].

## 5.3  Quantum Programming

Even if quantum computers are actually developed, we need to implement quantum algorithms for practical applications. Implementations are done by means of hardwares or softwares. For example, quantum gates enable quantum computing in hardware. But the use of quantum gates requires quite a lot of time and financial resources. This implies that hardware-based quantum computation is not always practical. If quantum computing is possible in the software level, the research and applications of quantum computing will be accelerated.

For this purpose, however, we need a special *programming language* for quantum computing. Fortunately, there are in fact several programming languages capable of quantum programming.

### 5.3.1  QCL

In 1998, Ömer at Technical University of Vienna implemented a first real quantum programming language called *QCL* (Quantum Computation Language); see Ömer [64]. Since then, many quantum programming languages have been developed. Now, QCL is available as a free software. Below we will briefly comment on QCL.

There are two advantages of using quantum programming languages. First, we can deal with quantum algorithms at the abstract level by means of quantum programming languages. Second, the formalization and simulation of quantum algorithms in quantum programming languages enable us to facilitate the implementation of different quantum architectures. In fact, these aspects contrast those in quantum computers like quantum gates.

Although these advantages hold for other (standard) programming languages, they seem to be essential to quantum computing. Ömer says in the preface of his Ph.D. thesis [64]:

> In contrast to quantum circuits, quantum Turing machines or the algebraic definition of unitary transformations, programming languages allow the complete and constructive description of quantum algorithms including their classical control structure for arbitrary input sizes and hardware architectures.

> This thesis investigates how the classical formalism of structured programming can be adapted to the field of quantum computing, based on the machine model of a universal computer with a quantum oracle allowing the application of unitary gates and the measurement of single qubits.

We believe that these remarks address the importance of quantum programming languages in connection with the research on quantum computing.

According to Ömer, any useful quantum programming language must have the following characteristics:

---

(1) it is constructive.

(2) it is hardware independent.

(3) it provides arbitrary level of abstraction.

(4) it integrates non-classical features at a semantic level.

---

(1) implies that a program can be written structurally. It is the fundamental basis for current programming in *structured programming* in that all programs are constructively presented.

(2) says that a program can be executed independent of the hardware on which it runs. This is useful to the portability of programs and the applicability of quantum computing can be enhanced. The feature cannot always be guaranteed in quantum gates.

By (3), we can consider an algorithm at an abstract level as well as a concrete level. This is important since quantum system can be described in various ways. Of course, the levels of abstraction depend on the problems in question.

(4) showed that we can describe a program for quantum mechanical operations, which are very different from the ones in standard programming languages. This is obviously the most important point in quantum programming languages.

The features (1)-(3) are applied to all programming languages, but (4) is a peculiarity of quantum programming languages. In other words, we can express quantum computation in the form similar to our conceptual level by means of quantum programming languages. This is surely attractive for quantum computing.

In fact, QCL has the peculiarities of quantum computing in addition to the standard structured programming. Ömer mentioned the following peculiarities:

---

- reversibility of unitary operations
- non-locality of qubits
- non-observavility of states
- destructive nature of measurement
- lack of an erase operation

---

Since unitary operations are reversible, quantum computation is also reversible. As addressed before, one of the important features of quantum computation is reversibility.

Qubits are not locally described, because non-locality is an important feature in existing quantum systems. And some quantum states may not be observables. Due to the uncertain principle, a measurement is destructive.

Erase is classical operations in the (classical) Turing Machine. But, the quantum Turing Machine does not support it.

Quantum programming corresponds to an abstract level of quantum computing, and it can be performed by quantum programming languages. To realize it in a traditional fashion, we need to combine structured programming with quantum programming.

Ömer calls such programming *structured quantum programming*. For this purpose, he implemented a quantum programming language, i.e., QCL. QCL is based on quantum mechanical counterparts on classical programming languages. The computational model for QCL is a quantum architecture instead of a Turing Machine (or Boolean algebra).

Quantum registers are the counterpart of variables. Quantum gates play a role of variable assignments. Classical input of data is performed by a measurement in the sense of quantum mechanics. The syntax of QCL is based on that of the context-free language, and it is similar to that of the C language. In addition, QCL has a simulator for testing. The details of QCL can be found in Ömer [64].

We believe that the study of quantum computing at the software level will be promoted in future. QCL can be regarded as initial work in this line. We expect to see quantum programming languages with more powerful programming features and programming environment.

## 5.3.2 Other Quantum Programming Languages

QCL is a first (real) quantum programming language, but several quantum programming languages have been also developed. We can list some of them as follows:

- Qgol
- qGCL
- Quantum C Language
- Q Language
- QPL
- Quantum Lambda Calculus

A comprehensive survey on quantum programming languages can be found in Gay [40].

### Qgol

*Qgol* is perhaps a first system for simulating quantum computation developed by Baker in 1996; see Baker [4]. Qgol can be seen as a simulator rather than a programming language. It is a pure quantum state manipulation language, inspired by functional programming.

It was designed as a visual language. The fundamental symbols of Qgol are gates (quantum operators). Qgol also has the graphical editor. Thus, Qgol is initial work for quantum programming language.

## qGCL
*qGCL* (quantum Guarded Command Language) was defined by Zuliani in 2001 based on Dijkstra's semantics for programming language; see Zuliani [88]. It thus has a formal semantics and an associated refinement calculus.

QGCL cannot only describe quantum algorithms but also provide a way for quantum program development and correctness proof. In this sense, we can say that qGCL has desirable features compatible to traditional programming theory.

## Q Language
*Q Language* was developed by Bettelli in 2002, which is viewed as an extension of C++; see Bettelli [12]. It has many classes for basic quantum operations and quantum registers, and the user can define new classes by means of its object-oriented features.

It is thus possible to interpret Q Language as an object-oriented programming language for quantum computing. This means that many futures of C++ can be used for quantum programming. Since it is based on C++, we need no special compliers for its uses. Q has been implemented with a simulator.

## QPL
*QPL* is designed by Selinger in 2004; see Selinger [71]. His approach adopts the slogan "quantum data, classical data". A program in QPL is written as a flowchart. QPL is functional in the sense of each statement operates by transforming a specific set of inputs to outputs.

It also provides high-level control features such as loops and recursion and structured data types such as trees and lists. It admits the o-called *denotational semantics*, which is one of the most formal semantics for programming languages.

QPL is thus a language which is classified as a functional programming language. Later, Mauerer proposed *cQPL* extending QPL with mechanisms for communication; see Mauerer [60].

## Quantum Lambda Calculus
*Quantum Lambda Calculus* was proposed by Maymin in 1996; see Maymin [61]. He defined two extensions of $\lambda$-calculus, i.e., $\lambda^p$-calculus (probabilistic $\lambda$-calculus) and $\lambda^q$-calculus (quantum $\lambda$-calculus). The latter can effectively simulate one-dimensional partitioned quantum cellular automata, which are equivalent to quantum Turing Machines.

Later, Van Tonder gave quantum $\lambda$-calculus with an operational semantics and an equational theory; see Van Tonder [81]. Quantum Lambda Calculus can be seen as a meta-language for functional programming language. It incorporates a type system to restrict the possible operations, and equational reasoning about can be done in it.

There are surely other quantum programming languages in the literature. We should also explore other types of quantum programming languages. For instance, a programming language based on quantum logic could be considered.

# Chapter 6
# Future of Quantum Computing

**Abstract.** Chapter 6 concludes the book with discussing the future of quantum computing. In particular, implementations and important problems of quantum computing are discussed.

## 6.1 Implementations of Quantum Computers

The final chapter prospects the future of quantum computing. First, we discuss implementations of quantum computers capable of running various quantum algorithms. To implement quantum computers is a great step for various applications of quantum computing. We focus on NMR quantum computers and trapped ion quantum computers. Other approaches are also briefly surveyed. Additionally, we discuss DiVincenzo's five criteria for general approaches to quantum computers.

Second, some problems of quantum computing are shown. These problems consist of theoretical and practical ones, and it is important to solve them to advance the research on quantum computing. Theoretical problems include: Artificial Intelligence, computability of quantum computers, and quantum logic and quantum computation. As practical problems, we discuss: general architecture of quantum computers, new quantum algorithms, and use of quantum programming languages. However, more work should be done for solving these problems.

Currently, theoretical foundations for quantum computing are well established for most aspects in quantum computing. In fact, several computational models, i.e., quantum Turing Machines and quantum gates, have been investigated in great detail by many researchers. In addition, many quantum algorithms, like Shor's algorithm and Grover's algorithm, which are still restricted, have been proposed. They actually addressed the importance and possibilities of quantum computers. It is thus clear that a next step is to develop quantum computers.

Unless quantum computers are actually used, various theories for quantum computing cannot be verified. In other words, we cannot benefit from quantum computers in our life. It is thus expected to see some quantum computers. But there are many difficulties with the implementation of quantum computers.

Now, we do not have a general architecture for quantum computers as the Neumann-type computer. In fact, there are several types of hardware of quantum computers in the literature. They have both advantages and disadvantages. The important point is that the hardware of a quantum computer has to implement qubits physically. However, we are faced with the problem of *decoherence*. By decoherence, we mean the phenomena in which a quantum state collapses affecting by the environment.

Particles have to interact with each other in quantum computing, but they influence each other's state. As a consequence, the qubits becomes decoherence, i.e., they hold only one value rather than multiple values. In fact, increasing qubits will cause more decoherence. But, for quantum computing, we provide initialization, control and measurement for qubits. Since a quantum computer is a physical device, qubits interact with the external world. Currently, NMR and trapped ion are considered to be attractive methods for implementing quantum computers. Thus, we will explain the basics of these methods in some detail.

### 6.1.1  NMR Quantum Computer

*NMR quantum computer* is a first approach to realize a quantum computer. *NMR* (Nuclear Magnetic Resonance) is the phenomena which happens when atomic nuclei are immersed in a static magnetic field and exposed to a second oscillating magnetic field.

Recently, NMR is applied to various areas. For instance, in medicine, the so-called *MRI* (Magnetic Resonance Imaging) is an imaging technique to produce high quality images of the inside of the human body, and it is becoming an alternative to the X-ray diagnosis. It is also applied to chemical analysis, material science, and others.

As described in Chapter 3, a nucleus and particles like an electron have a *spin* which is an inherent physical value. The spin of a nucleus interacts with the spin of electrons, i.e., the *electron spin* surrounding it. When a nucleus is in magnetic a field, some of the protons therein line up in parallel with the field and some line up against it. This shows that a nucleus spine has only two states, i.e., parallel and anti-parallel states.

Let the parallel state be $|0\rangle$ and the anti-parallel be $|1\rangle$. Then, we can construct a qubit as illustrated in Figure 6.1

| parallel | ↑↑  ↓↓ |
| anti-parallel | ↑↓  ↓↑ |

**Fig. 6.1** Parallel and anti-parallel states

In 1998, Gershenfeld and Chuang at MIT succeeded in developing a 2-qubits quantum computer; see Gershenfeld and Chuang [41]. It is a first real quantum computer based on bulk spin-resonance. They say:

One great advantage of this approach is that, because of the massive redundancy provided by having a large ensemble of identical copies of the system, environmental interactions or intentional measurements only weakly perturb the computer's state. Thus, quantum computation becomes experimentally accessible in many naturally existing materials.

In fact, their work was a first step towards the implementation of quantum computers whose significance is, as the quotation shows, lies in the fact that quantum computers can be implemented by existing stuffs, i.e., NMR and material, i.e., molecules.

In their approach, a molecule can then be seen as a single computer, whose state is determined by the orientation of its spins, i.e., parallel and anti-parallel. Sequences of radio frequency pulses for manipulating spins constitute quantum logic gates and perform unitary transformations in the state. This enables to realize some qubits quantum computer. Generally, in NMR, a signal can be obtained by measuring the ensemble average of about $10^{18}$ molecules. It is thus considered that such a measurement does not affect the state of a nucleus spin.

But, there is a theoretical limitation that we can maximally construct about 7-qubits computers in this approach since the polarization is small. In 2001, IBM developed 7-qubits NMR quantum computer and implemented Shor's algorithm; see Vandersypen et al. [80]. IBM is now continuing the research on NMR quantum computer.

NMR quantum computers have the advantage that they can be constructed in the environment with normal temperature. In contrast, other methods like trapped ion quantum computer requires a special environment. But, to implement practical quantum computers in this method, we need a new technique to detect spins and control decoherence.

## 6.1.2 Trapped Ion Quantum Computer

*Trapped ion quantum computer* is an alternative approach to quantum computers. It is a device capable of locking up charged particles like an electron into a closed space. Trapping ions with electromagnetic fields in a very low temperature and in vacuum electromagnetic field, we can stop ions in a chain. The spin of free electors in ions separated by the trapped ion can be manipulated by laser light. Since the spin of an electron is $\pm\frac{1}{2}$, we can construct qubits by means of trapped ions.

Trapping the ions with electromagnetic forces in vacuum isolates them almost perfectly from their environment. Thus, it is possible to store quantum information by trapped ions. Additionally, the internal states can be initialized and measured with high accuracy.

In 1995, Cirac and Zoller developed a trapped ion quantum computer and realized the CNot gate; see Cirac and Zoller [21]. Their idea is to use laser pulses to give an interaction of the electronic states of individual ions. Then, quantum computation can be performed by coupling the ion with 2-qubits. They also proved that the size of the resources necessary to control trapped ions does not increase exponentially with the number of qubits.

Kielpinski et al. [55] described a general architecture for ion trapped quantum computer. However, trapped ion quantum computers need a large scaled device, i.e., trapped ion giving a very low temperature and vacuum. It is not easy to build such a device. The approach also faces the difficulty with the construction of many qubits. One solution is to use quantum communication to link a number of small ion trapped quantum computers.

### 6.1.3  Other Approaches

There are other approaches to the implementation of quantum computers. In fact, the hardwares based on quantum dot, Josephson junction, and photon have been investigated. As we showed the idea of the third approaches, we below simply discuss the first two approaches.

**Quantum Dots**

A *quantum dot* (QD) is the state of an electron locked up in a three-dimensional space, which can be seen as tiny particles, or nanoparticles. Since a quantum dot can be constructed in a box consisting of semiconductors or metals, it is considered to be easy to make a hardware by means of quantum dots.

It is possible to construct small quantum gates by using quantum dots. Quantum dots are nanoscale structures built in semiconducting materials that hold tiny puddles of electrons, giving each dot the spin. Since the spins of dots are either up or down, they can form qubits. Several models for realizing quantum dot based quantum computers have been proposed in the literature.

Loss and DiVincenzo proposed a new implementation of a quantum computer based on the spin states of coupled single-electron quantum dots in 1998; see Loss and DiVincenzo [59]. In their model, the qubit is realized as the spin of the excess electron on a single-electron quantum dot. Based on the idea, they described a universal set of one- and two-qubit quantum gates.

Kane also showed a different model for a quantum computer using an array of nuclear spins located on donors in silicon, the semiconductor used in most conventional computer electronics; see Kane [53]. Logical operations and measurements can in principle be performed independently and in parallel on each spin in the array.

**Josephson Junction**

*Josephson junction* is another promising base for quantum computers. It consists of a cylindrical substrate having a flat end face disposed in a plane perpendicular to the axis of the cylinder based on the so-called *Josephson effect*, which is the phenomena of supercurrent, i.e., flow of electric current between two pieces of superconducting material separated by a thin layer of insulating material. Josephson effect was predicted by Josephson in 1962; see Josephson [51].

Josephson junctions can be viewed as a macroscopic quantum system, in that they correspond to qubits, and can be applied to quantum computers; see Averin [3]. They are, as quantum dots, a promising basis for scalable quantum computers, but its cost becomes very high since the environment with a very low temperature is required.

As reviewed above, there are several approaches to develop quantum computes. Now, it is possible to realize quantum computers with more than 10-qubits, and the research will be progressed.

### 6.1.4 DiVincenzo's Five Criteria

Today, there have no general architectures for implementing quantum computers, but *DiVincenzo's five criteria* is known; see DiVincenzo [31]. DiVincenzo proposed the following five criteria, and later expanded to seven criteria, to realize a quantum computer.

---

**DiVincenzo's five criteria**

(C1) A scalable physical system with well characterized qubits

(C2) The ability to initialize the state of the qubits to a simple fiducial state

(C3) Long relevant decoherence times, much longer than the gate operation time

(C4) A "universal" set of quantum gates

(C5) A qubit-specific measurement capability

---

These criteria are interpreted as the necessary conditions saying that certain physical systems are the base for quantum computers.

(C1) means that the number of qubits is enough and can be increased. More concretely, we need a physical system containing a collection of qubits. A well characterized qubit can be interpreted in such a way that its physical parameters should be accurately known.

There are some ways to give well characterized qubits. For example, a two-state system is one of the most basic characterizations. The maximum number of current quantum computers is about 7-qubit, and 1-qubit or 2-qubits are generally used. Thus, it seems to be important to have a technology for connecting qubits.

(C2) refers to the ability to initialize qubits. For quantum computing, it is necessary to initialize a qubit to the fiducial states like $|00\cdots0\rangle$. This is nothing but the initializations of quantum registers. Initializing quantum registers with a known value is necessary before the start of computation. The requirement is also important to quantum error correction.

There are two main approaches to such an initialization. One is to naturally cool the system when the ground state of its Hamiltonian is the state of interest. The other is that the standard state can be achieved by a measurement which projects the system either into the state desired or another state which can be rotated into it. However, these approaches are not fundamentally different from one another.

(C3) states that decoherence does not happen for certain period, namely the so-called *decoherence time* is needed. Decoherence time characterizes the dynamics of a qubit in contact with its environment. Decoherence is very important to the implementation of quantum computers, because it is identified with the principal mechanism for the emergence of classical behavior. At the same time, it implies that decoherence is dangerous for quantum computing.

If the decoherence time is very long, then the capability of the quantum computer cannot exceed that of classical one. This means that decoherence time must be long enough. For quantum computation, we have to decrease decoherence by the inter-action with the external world by stabilizing the quantum states until a computation ends. For this purpose, decoherence times are needed to guarantee the stable states during computation. One solution to the problem of decoherence times appears to incorporate a technology of quantum error correction.

(C4) means that we need fundamental quantum gates for a quantum computer. It can be paraphrased that arbitrary quantum gates can be constructed. Of course, it is the basic requirement of quantum computers. As described in Chapter 4, quantum gates can be specified as the unitary transformation. And it was proved that there exist some universal quantum gates.

We can describe a quantum algorithm by a sequence of unitary transformations. Each unitary transformation is applied to several qubits which are typically no more than three. It is, however, very difficult to implement quantum gates perfectly. In fact, there are systematic and random errors in the implementation. They are considered as another source of decoherence. Thus, effective error corrections are required for reliable computation.

(C5) refers to the ability to read out the result of a computation after a measurement of specific qubits. In other words, if we can correctly measure qubits after a computation, the precision of results can be improved. Of course, it is desirable to have a measurement with 100 % quantum efficiency (i.e., precision), but there is a trade-off between quantum and computational efficiency.

For the practical level, it seems that we require more than 90% quantum efficiency. A measurement should be completed efficiently. If so, then its repeated application in quantum computation is useful for simplifying quantum error correction. But, if a measurement is not fast, we need to implement a number of quantum gates for quantum error correction.

Later, DiVincenzo pointed out that we need additional two criteria for quantum communications in DiVincenzo [32]. They are needed if quantum computers deal with quantum communication. It is obvious that they are closely related.

---

(C6) The ability to interconvert stationary and flying qubits

(C7) The ability faithfully to transmit flying qubits between specified locations

---

(C6) concerns the ability of interconnecting stationary qubits used in local quantum computer and flying qubits used in quantum communications. This implies that various transformations of qubits are necessary.

(C7) asserts that the reliability of communication of qubits is high. In other words, this is to be able to transmit flying qubits correctly. The requirement is indispensable to realize such quantum technologies as quantum code and quantum teleportation.

We believe that DiVincenzo's five (or seven) criteria is of special importance to the promotion and evaluation of various approaches to quantum computers. Although it may be far from complete, it provides general criteria for quantum computers. DiVincenzo says in [32]

> No one can see how or whether all the requirements above can be fulfilled, or whether there are new tradeoffs, not envisioned in our present theoretical discussions but suggested by further experiments, that might take our investigations in an entirely new path.

where all the requirements refer to his criteria. As he mentioned here, we may be able to obtain a new possibility of the implementation of quantum computers in the near future.

In our opinion, in the 2020's, some special purpose quantum computers will be realized. They are designed to solve specific problems, e.g., scientific problems. They will be incorporated into standard computer as a coprocessor. We believe that in the 2030's general purpose quantum computers will emerge and quantum computing will become a core of information technology. Of course, we hope that our expectation comes off and we expect to see the realization of quantum computers early.

But, the realization of quantum computers will raise new problems which are undoubtedly independent of technology. The development of a quantum computer clearly needs extraordinary cost. If we have a quantum computer, this means that we can use a *great* technology. We can say that quantum computers can give a technology compatible to nuclear energy.

For instance, the nations which establish the technology of quantum computing will dominate politically as well as military. This is because quantum computing plays an important role. Due to such a possible situation, it may be necessary to regulate the research on quantum computers at the state level. In fact, the U.S. government started with the discussion on export control on quantum computers in 2005. If it is materialized, then the research on quantum computing will be naturally controlled. This means that its development will certainly affect.

It also goes without saying that international research collaboration is needed to realize quantum computers, since the development of quantum computers need extraordinary cost and many researcher (or engineers). However, there is an issue open to question: Do politicians know qubits?

## 6.2    Problems of Quantum Computing

Finally, we raise several problems in quantum computing. Of course, there are many important problems. But, we discuss not all interesting problems here. Below, we argue some of theoretical problems and practical problems separately.

The theoretical problems in question are as follows:

---

(1) Artificial Intelligence and quantum computation

(2) Computability of quantum computers

(3) Quantum logic and quantum computation

---

These problems appear to challenge many theoreticians working on quantum computing as well as other related areas.

Next, we argue the following practical problems:

---

(1) General architecture of quantum computers

(2) New quantum algorithms

(3) Use of quantum programming languages

---

To solve these problems is of special importance to applications of quantum computer in the near future.

### 6.2.1    Theoretical Problems

The first problem is about the connection of Artificial Intelligence and quantum computers. By the Church-Turing principle in theory of computation, every computer is functionally equivalent. This implies that certain computer can be simulated by other computers.

It is thus natural to consider that our thought can be simulated by a quantum computer. So far, our intelligent activities in connection with computing have been studied in the area called *Artificial Intelligence* (AI). One of the purposes of Artificial Intelligence is to realize a computer which can think. But existing approaches to Artificial Intelligence are generally faced with the problem of vast computational cost. To solve the problem, parallel computing is used to speed up the computation needed for Artificial Intelligence.

The development of quantum computers can be regarded as a key to the new phase of Artificial Intelligence, because one of the striking features of a quantum computer is to perform ultra high speed computation. It means that a quantum computer can efficiently compute AI algorithms which are intractable in classical computers.

But, on the other hand, there are indications saying that human's consciousness is mysterious and another physical principle is necessary, as Penrose pointed out. It is a point of view that Artificial Intelligence cannot be realized even if quantum computers are developed.

In 1994, Penrose proposed *Quantum Mechanisms of Consciousness*, which unifies quantum mechanics and the theory of relativity. He also argued that it is necessary to realize Artificial Intelligence; see Penrose [66] and Hameroff and Penrose [45].

**Roger Penrose (U.K.: 1931-)**

Roger Penrose was born in Colchester, U.K. on 1931. He studied University College School and University College, London. He received Ph.D. degree from St John's College at Cambridge in 1958. He contributed to mathematics and theoretical physics. For example, in 1965, he proved that every black hole contains a singularity. Penrose wrote several books discussing the connection of physics and human consciousness, notably, *The Emperor's New Mind* (1989), *Shadows of the Mind* (1994), and *The Road to Reality* (2005). Penrose continues to develop quantum mechanical approaches to human's consciousness as opposed to Artificial Intelligence.

Quantum Mechanisms of Consciousness assumes that the principles of quantum mechanics are also valid for the macro levels like our brain. So far, quantum mechanics has been investigated as a theory which deals with the micro levels like atom and particle, and has not been used for studying the macro levels.

Penrose applied it to formalize the behaviors of the brain which is the origin of our thought. This is a starting point of Quantum Mechanics of Consciousness. Hameroff and Penrose [45] say:

> If conscious experience is intimately connected with the very physics underlying space-time structure, then **Orch OR** in microtubules indeed provides us with a completely new and uniquely promising perspective on the hard problem of consciousness.

where Orch OR refers to the orchestrated (i.e., self-organized) objective reduction. Since the details of their theory is beyond the scope of this book, we do not proceed further here.

It is an important problem whether quantum computing is of help to the study of Artificial Intelligence. We should also discuss in detail whether theories like Quantum Mechanics of Consciousness are needed to the realization of Artificial Intelligence.

The second problem is concerned with the problem of computability of quantum computers. We know that a quantum Turing Machine and a probabilistic Turing Machine are computationally equivalent as discussed before. But this means that a quantum Turing Machine cannot solve the problems which cannot be solved by classical Turing Machine. From this, opponents may claim that we do not need quantum computers. Therefore, it is necessary to justify the computable power of quantum Turing Machines more precisely. Because classical Turing Machine is also a (classical) physical system, the computational power of quantum Turing Machines appears to be higher than that of classical Turing Machine.

To address the problem, we may work out a new formalization of Turing Machine and provide more strict interpretations of Church-Turing-Deutsch principle. However, there are no adequate formalizations. If we can prove that quantum computers are more powerful than classical computers in computational power under *certain* interpretations and conditions, then it is theoretically very interesting.

If quantum computer is beyond of standard computer, then the possibility that quantum computers can contribute to the research on Artificial Intelligence will increase. In other words, quantum computing can serve as a viable framework for Artificial Intelligence. Even if we obtain the usefulness of quantum computers for Artificial Intelligence, we must face a difficulty, however. It is the problem of *heuristic*, which is a technique for solving a problem more efficiently. Undoubtedly, human seem to use various heuristics for problem solving. For example, common-sense reasoning can be done by jumping to conclusions using some heuristics. In fact, most algorithms used in Artificial Intelligence should be given algorithmically as well as heuristically.

The third problem is about the status of the so-called *quantum logic* in quantum computing. As is well known, quantum logic was proposed by von Neumann and refers to the family of logical systems based the Hilbert space; see Birkhoff and von Neumann [14] and Dalla Chiara [22]. In 1936, Birkhoff and von Neumann formalized quantum logic as a logical system for projection operators on the Hilbert space. A proposition of quantum logic represents quantum mechanical phenomenon after a measurement. Because quantum logic is different from classical logic, it is classified as non-classical logic.

**Garrett Birkhoff (USA: 1911-1996)**

Garrett Birkhoff was born in Princeton, USA on 1911. He entered Harvard University in 1928 and received M.A. in 1932. He was appointed as an instructor at Harvard in 1936. In 1969, he became George Putnum Professor of Pure and Applied Mathematics at Harvard. He is well known for his contributions to universal algebra and lattice theory. In 1936, he proposed quantum for quantum mechanics, with von Neumann. His book *Lattice Theory* (1940) is standard textbook on lattice theory. Birkhoff died in Water Mill, USA on 1996.

One of the striking features of quantum logic is the point that the distributive law, which holds in classical logic:

$$P \wedge (Q \vee R) \leftrightarrow (P \wedge Q) \vee (P \wedge R)$$

does no hold. Consequently, the semantics for quantum logic requires an algebraic structure different from the Boolean algebra for classical logic. There is a rich literature on quantum logic; see Dalla Chiara [22].

However, quantum logic has been criticized by many people. For example, Bub discussed philosophical aspects of the interpretation of quantum mechanics in connection with quantum logic; see Bub [18]. Thus, philosophical and logical investigations of quantum logic should be continued.

If the motivation that quantum logic is a logic for quantum mechanics is correct, quantum logic should be also applied to quantum computing. This is because both quantum logic and quantum computing are grounded on the notion of the Hilbert space. It can be pointed out that quantum logic has a proof theory which could be interpreted as a computational mechanism.

There appear to be some connections of quantum logic and quantum computer. But they are not obvious. One possible connection is, as mentioned above, to simulate quantum computation within a quantum logical calculus. Other possible connections useful to quantum connection could be obtained in algebraic setting. This is because the foundation of quantum logic has been described based on special form of algebra. However, little work has been done focusing on the connections since quantum logic has not been formalized as the basis for *computation*.

## 6.2.2   Practical Problems

Finally, we consider some practical problems related to quantum computers. As discussed in the previous section, many people are making efforts to implement quantum computers in various ways. Many quantum algorithms have been proposed in the literature. But, we must focus on practical problems.

The first problem is to obtain a general architecture like the universal quantum computer. In fact, most approaches to quantum computers adopt special methods to give qubits. But, it is not obvious which architecture is effective for the implementation of quantum computers.

In this regard, DiVincenzo's criteria as discussed above will be of help to establish such a general architecture for quantum computers. We think that all criteria may be important and they have difficulties. But, one of the most difficult problem is to discover quantum mechanical phenomena which can be easily implemented as qubits.

The second problem is the study of *new* quantum algorithms. In fact, various fast algorithms have been devised for classical computer, and they form the basis of current computer applications. Of course, many new quantum algorithms were proposed for quantum computing. But, we have the case that the algorithms like Shor's algorithm, which were not used before, are suited to quantum algorithms. From this observation, foundational thinking of quantum algorithm and the effective transformations of classical algorithms into quantum algorithms are to be investigated.

It has been often claimed that in current theories of quantum computing quantum computers are able to solve certain problems very faster than classical computers do. In fact, this is an advantage of quantum computers over current commuters. Defenders of quantum computers address the point.

We think that the identification of the problems which are suited to quantum algorithms can clarify the promising areas in quantum computing. Such identification is important in that it can enhance the applicability of quantum computing. It is, however, difficult to identify such problems because of the nature of quantum algorithms.

The third problem is the effective use of quantum programming languages. By considering the difficulties in the implementations of quantum computers, it may be attractive to employ quantum programming languages to implement various quantum algorithms in the software level. Additionally, we have to establish their programming techniques. Also noted is that quantum programming languages like QPL can simulate quantum computing at various levels.

Here, the important point is that such simulations are independent to hardware. Namely, we can write a quantum program which does not depend on specific quantum computer. This will enable us to facilitate the evaluation and comparison of theories of quantum computing at the software level instead of the hardware level. It is thus possible to say that the use of quantum programming languages is useful to the study of quantum computing.

But, for quantum programming, we need to have wide range of knowledge on programming, quantum mechanics and quantum computing. In other words, one should learn various things which are needed to understand quantum computing. However, it is not a trivial task to do it. To develop practical quantum programs, one need to learn more, e.g., design and verification. It may not be easy for existing programmers to learn quantum computing for many reasons. Thus, the number of skillful programmers who can contribute to quantum computing cannot increase rapidly.

To overcome the problem, it is necessary to illuminate the usefulness of quantum computing. It is also expected that the relevant educations are given at educational institutions like university. Most computer science departments in university do not always teach quantum mechanics. Similarly, the education of computer science at physics departments is not enough.

It is of course important to give a comprehensive review of quantum computers so that ordinary people are interested in them. It is now not true that quantum computer is not a familiar thing. Thus, we have to give understandable reviews for quantum computer, but it is not easy to do it. We also expect see good textbooks on quantum computing; see Nielsen and Chuang [63].

As mentioned above, although there are many negative factors, we believe that quantum computers will be realized and quantum computing will become one of the important technologies in computer science within twenty years. Of course, many people want to contribute to the field of quantum computing now. However, only a few can do. This is a problem in the area. There are many problems. One of the most serious problems lies in the status of researchers working on quantum computing.

At present, we feel that the background of many researchers for quantum computing seems to be physics. In fact, those who invented the principle of quantum computers and main quantum algorithms are also physicists. It is not surprising to see the facts. We hope that many people working on computer science also study quantum computing and those who are in physics also study computer science.

Twenty years ago, the theory and technology of current computer were very difficult for most people. In other words, every people could not always properly understand them and only experts could obtain them. However, we can say that they are now commonsense for many people.

Several useful textbooks and monographs on quantum computing will be written by experts. Thus, we believe that the contents of this book, i.e., knowledge of quantum computing, will be commonsense twenty years later. And we will be able to revise the book in the near future.

When we were completing this book (December 2012), we encountered a big news related to quantum computing. Serge Haroche (France) and David J. Wineland (USA) received the Nobel Prize in Physics for 2012. Some people was surprised to have it.

They in fact contributed to ground-breaking experimental methods that enable measuring and manipulation of individual quantum systems. They demonstrated the methods for direct observation of individual quantum systems without destroying them. The methods are clearly considered to be useful to the implementation of quantum computers; see Haroche [46] and Wineland [85].

In 7 October 2012, the Royal Swedish Academy of Sciences informed the reason of the decision, which says:

> Their ground-breaking methods have enabled this field of research to take the very first steps towards building a new type of super fast computer based on quantum physics. Perhaps the quantum computer will change our everyday lives in this century in the same radical way as the classical computer did in the last century.
> (http://www.nobelprize.org/nobel_prizes/physics/laureates/2012)

Here, we can understand that "this field" obviously means quantum mechanics and that "a new type of super fast computer" a quantum computer (specifically trapped ion quantum computer).

The breaking news reveals that quantum computing is now recognized as an important subfield of physics. However, the reason should be rewritten for us now. Although the Royal Swedish Academy of Sciences uses the term "perhaps" as above, we can say as follows:

> Quantum computers will be able to change our everyday lives in this century in the same radical way as the classical computer did in the last century.

which is, we believe, undoubtedly one of the reasons that we must (or want to) study quantum computing.

# References

1. Aharonov, D., van Dam, W., Kempe, J., Landau, Z., Lloyd, S., Regev, O.: Adiabatic quantum computation is equivalent to standard quantum computation. In: Proc. of the 45th Annual IEEE Symposium on Foundations of Computer Science (FOCS), pp. 42–51 (2004)
2. Ambainis, A.: Quantum walk algorithm for element distinctness. SIAM Journal of Computing 37, 210–239 (2007)
3. Averin, D.: Adiabatic quantum computation with Cooper pairs. Solid State Communication 105, 659–664 (1998)
4. Baker, G.: Qgol. Honors Degree, Department of Computing, Macquarie University (1996)
5. Bell, J.: On the Einstein Pdolsky Rosen paradox. Physics 1, 195–200 (1964)
6. Benioff, P.: The computer as a physical system. Journal of Statistical Physics 22, 563–591 (1980)
7. Bennett, C.: Logical reversibility of computation. IBM Journal of Research and Development 17, 525–532 (1973)
8. Bennett, C.: Thermodynamics of computation-a review. International Journal of Theoretical Physics 21, 905–940 (1982)
9. Bennett, C., Brassard, G., Ekert, A.: Quantum cryptography. Scientific American 267, 50–57 (1992)
10. Bennett, C., Brassard, G., Crepeau, C., Josza, R., Peres, A., Wooters, W.: Teleporting an unknown quantum state via dual classical and Einstein-Podolsky-Rosen channels. Physical Review Letters 70, 1895–1899 (1993)
11. Bernstein, E., Vazirani, U.: Quantum complexity theory. In: Proc. of the 25th Annual Symposium on the Theory of Computing, pp. 11–20 (1993)
12. Bettelli, S.: Toward an Architecture for Quantum Programming. University of Toronto, Ph.D. thesis (2002)
13. Birkhoff, G.: Lattice Theory. American Mathematical Society (1940)
14. Birkhoff, G., von Neumann, J.: The logic of quantum mechanics. Annals of Mathematics 37, 823–843 (1936)
15. Boole, G.: The Mathematical Analysis of Logic. MacMillan, Barclay & MacMillan, Cambridge (1847)
16. Boolos, G., Jeffrey, R.: Computability and Logic, 3rd edn. Cambridge University Press, Cambridge (1989)
17. Brouwmeester, D., Pan, J., Mattle, K., Eibl, M., Weinfurter, H., Zielinger, A.: Experimental quantum teleportation. Nature 390, 575–579 (1997)

18. Bub, J.: The Interpretation of Quantum Mechanics. Reidel, Dordrecht (1974)
19. Church, A.: An unsolvable problem of elementary number theory. American Journal of Mathematics 58, 345–366 (1936)
20. Church, A.: The Calculi of Lambda-Conversion. Princeton University Press, Princeton (1941)
21. Cirac, J., Zoller, P.: Quantum computation with cold trapped ions. Physical Review Letters 74, 4091–4094 (1995)
22. Dalla Chiara, M.: Quantum logic. In: Gabbay, D., Guenthner, F. (eds.) Handbook of Philosophical Logic, vol. III, pp. 427–469. Reidel, Dordrecht (1986)
23. Davey, B.A., Priestley, H.A.: Introduction to Lattices and Order, 2nd edn. Cambridge University Press, Cambridge (2002)
24. Deutsch, D.: Quantum theory, the Church-Turing principle and the universal quantum computer. Proc. of the Royal Society A 400, 97–117 (1985)
25. Deutsch, D.: Quantum computational networks. Proc. of the Royal Society A 425, 73–90 (1989)
26. Deutsch, D., Barenco, A., Ekert, A.: Universality in quantum computation. Proc. of the Royal Society A 449, 669–677 (1995)
27. Deutsch, D., Jozsa, R.: Rapid solution of problems by quantum computation. Proc. of the Royal Society A 439, 553–558 (1992)
28. Deutsch, D., Ekert, A., Lupacchini, R.: Machine, logic and quantum physics. Bulletin of Symbolic Logic 6, 265–283 (2000)
29. Dirac, P.: Principles of Quantum Mechanics, 4th edn. Clarendon Press, Clarendon (1982)
30. Dirac, P.: The physical interpretation of quantum mechanics. Proc. of the Royal Society A 180, 1–39 (1942)
31. DiVincenzo, D.: Quantum computation. Science 270, 255–261 (1995)
32. DiVincenzo, D.: The physical implementation of quantum computation. Fortschritte Der Physik 48, 771–783 (2000)
33. Einstein, A., Podolsky, B., Rosen, N.: Can quantum-mechanical description of physical reality be considered complete? Physical Review 47, 777–780 (1935)
34. Farhi, J., Goldsone, J., Gurmann, S., Lapan, J., Lundgren, A., Preda, A.: A quantum adiabatic evolution algorithm applied to random instances of an NP-complete problem. Science 292, 472–476 (2001)
35. Farhi, J., Gutmann, S.: Quantum computation and decision trees. Physical Review A 58, 915–928 (1998)
36. Feynman, R.: Simulating physics with computers. International Journal of Theoretical Physics 21, 467–488 (1982)
37. Feynman, R.: Quantum mechanical computer. Optics News 11, 11–20 (1985)
38. Feynman, R., Hey, A., Allen, R.: Feynman Lectures on Computation. Perseus Books, Reading (1996)
39. Fredkin, E., Toffoli, T.: Conservative logic. International Journal of Theoretical Physics 21, 219–253 (1982)
40. Gay, S.: Quantum programming language: survey and bibliography. Mathematical Structures in Computer Science 16, 581–600 (2006)
41. Gershenfeld, N., Chuang, I.: Bulk spin resonance quantum computing. Science 275, 350–356 (1997)
42. Green, H.: Matrix Mechanics. P. Noordhoff, Groningen (1964)
43. Grier, D.A.: The ENIAC, the verb "to program" and the emergence of digital computers. Annals of the History of Computing, 51–55 (1996)
44. Grover, L.: A fast quantum mechanical algorithm for database search. In: Proc. of the 28th ACM Symposium on Theory of Computing, pp. 212–219 (1996)

45. Hameroff, S., Penrose, H.: Conscious events as orchestrated space-time selections. Journal of Consciousness Studies 3, 35–53 (1996)
46. Haroche, S.: Controlling photons in a box and exploring the quantum to classical boundary. Nobel Lecture (2012)
47. Heisenberg, W.: Über quantentheoreische Umdeutung kinematischer unde mechanischer Besiehungen. Zeitschrift für Physik 33, 879–893 (1925)
48. Heisenberg, W.: Über den anschaulichen Inhalt de quantentheoreische Kinematik undv Mechank. Zeitschrift für Physik 43, 172–198 (1927)
49. Holevo, A.: Some estimates for the amount of information transmittable by quantum communication channel. Problems of Information Transmission 9, 177–183 (1973) (in Russian)
50. Holevo, A.: The capacity of quantum channel with general signal states. IEEE Transactions on Information Theory 44, 269–273 (1998)
51. Josephson, B.: Possible new effects in superconductive tunneling. Physics Letters 1, 251–253 (1962)
52. Jozsa, R.: Quantum factoring, discrete logarithms, and the hidden subgroup problem. Computing in Science and Engineering 3, 34–43 (2001)
53. Kane, B.: A silicon-based nuclear spin quantum computer. Nature 393, 133–137 (1998)
54. Kempe, J.: Quantum random walk-An introductory overview. Contemporary Physics 44, 302–327 (2003)
55. Kielinski, D., Monroe, C., Wineland, D.J.: Architecture for a large-scale ion-trap quantum computer. Nature 417, 709–711 (2002)
56. Kleene, S.: Introduction to Metamathematics. North-Holland, Amsterdam (1952)
57. Landauer, R.: Irreversibility and heat generation in the computing process. IBM Journal of Research and Development 5, 183–191 (1961)
58. Lee, N., Benichi, H., Takeno, Y., Takeda, S., Webb, J., Huntington, E., Furusawa, A.: Teleportation of nonclassical wave packets of light. Science 332, 330–333 (2011)
59. Loss, D., DiVincenzo, D.: Quantum computation with quantum dots. Physical Review A 57, 120–126 (1998)
60. Mauerer, W.: Semantics and Simulation of Communication in Quantum Computing. Master thesis, University of Erlagen-Nuremberg (2005)
61. Maymin, P.: Extending the lambda calculus to express randomized and quantumized algorithms. arXiv:quant-ph/9612052 (1997)
62. Mendelson, E.: Introduction to Boolean Algebra and Switching Circuits. McGraw-Hill, New York (1973)
63. Nielsen, M., Chuang, I.: Quantum Computation and Quantum Information. Cambridge University Press, Cambridge (2000)
64. Ömer, B.: Structured Quantum Programming, Institute for Theoretical Physics, Vienna University of Technology. Ph.D. dissertation (2003)
65. Pauli, W.: Zur Quantenmechanik des magnetischen Elektrons. Zeitschrift für Physik 43, 601–623 (1927)
66. Penrose, R.: Shadows of the Mind. Oxford University Press, Oxford (1994)
67. Rivest, R., Shamir, A., Adleman, L.: A method for obtaining digital signatures and public-key cryptosystems. Communications of the ACM 21, 120–126 (1978)
68. Shannon, C.E.: A mathematical theory of communication. Bell System Technical Journal 27, 379-423, 623-656 (1948)
69. Shannon, C.E.: Communication theory of secrecy systems. Bell System Technical Journal 28, 656-715, 623-656 (1949)
70. Schrödinger, E.: Die gegenwärtige Situation in der Quantenmechanik. Naturwissenschaften 23, 807-812, 823-828, 844-849 (1935)

71. Selinger, P.: Towards a quantum programming language. Mathematical Structures in Computer Science 14, 527–586 (2004)
72. Shor, P.: Algorithms for quantum computation: Discrete logarithms and factoring. In: Proc. of the 35th Annual Symposium on Foundations of Computer Science, pp. 124–134 (1994)
73. Shor, P.: Polynomial-time algorithms for prime factorization and discrete logarithms on a quantum computer. SIAM Journal of Computing 26, 1484–1509 (1997), Expanded version of [72]
74. Shor, P.: Scheme for reducing decoherence in quantum computer memory. Physical Review A 52, 2493–2496 (1995)
75. Simon, D.: On the power of quantum computation. In: Proc. of the 35th Annual Symposium on Foundations of Computer Science, pp. 116–123 (1994)
76. Steane, A.: Error correcting codes in quantum theory. Physical Review Letters 77, 793–797 (1996)
77. Strang, G.: Introduction to Linear Algebra, Fourth Edition. Wellesley-Cambridge Press, Wellesley (2009)
78. Toffoli, T.: Bicontinuous extensions of invertible combinatorial functions. Mathematical Systems Theory 14, 13–23 (1981)
79. Turing, A.: On computable numbers, with an application to the Entscheidungproblem. Proc. of the London Mathematical Society 42, 230–265 (1936)
80. Vandersypen, L., Steffen, M., Breyta, G., Yannni, C., Sherwood, M., Chuang, I.: Experimental realization of Shor's quantum factoring algorithm using nuclear magnetic resonance. Nature 414, 883–887 (2001)
81. van Tonder, A.: A lambda calculus for quantum computing. SIAM Journal of Computing 33, 1109–1135 (2004)
82. von Neumann, J.: Mathematische Grundlagen der Quantenmechanik. Springer, Berlin (1932)
83. von Neumann, J.: First Draft of a Report to the EDVAC. University of Pennsylvania (1945)
84. Wiesner, S.: Conjugate coding. SIGACT News 15, 78–88 (1983)
85. Wineland, D.J.: Superposition, entanglement, and raising Schroedinger's cat. Nobel Lecture (2012)
86. Wootters, W., Zurek, W.: A single quantum cannot be cloned. Nature 299, 802–803 (1982)
87. Young, N.: An Introduction to Hilbert Space. Cambridge University Press, Cambridge (1988)
88. Zuliani: Quantum Programming. Ph.D. thesis, University of Oxford (2001)

# Glossary

**Artificial Intelligence (AI).** A subfield of computer science which aims to automate human intelligence.

**Atom.** A basic unit of matter. It consists of a nucleus encircled by moving electrons.

**Axiomatic System.** A formulation of logical system by set of axioms and rules of inference.

**Bit.** A basic unit of information in classical computers.

**Boolean Algebra.** An algebra invented by Boole in the 19th century to calculate human thinking. It is a basis of logical gates used in modern computers.

**Bra-Ket Notation.** A concise and convenient notation to describe quantum states introduced by Dirac in 1930.

**CCNot.** Quantum gate (controlled-controlled NOT). It is also called Toffoli gate.

**Church's Thesis.** A thesis that a computable function is Turing computable function. It was proposed by Church in 1935 and is also called Church-Turing thesis.

**Church-Turing Principle.** A thesis that every finitely realizable physical system can be perfectly simulated by a universal model computing machine operating by a finite means. It was proposed by Deutsch in 1985.

**Classical Physics.** Physics of matters in the macro level including Newtonian mechanics and electromagnetism.

**CNot.** Quantum gate (controlled NOT).

**Compton Effect.** The phenomena discovered by Compton in 1933, convincing that light is made up of particles.

**Copenhagen Interpretation.** The principle to explain the wave-particle duality in terms of a *collapse* of the wave function defended by Bohr.

**Decoherence.** The mechanism of quantum systems to interact with their environments to show probabilistically additive behavior. It is also called quantum decoherence.

**DiVincenzo's Five Criteria.** The requirements of the physical implementation of quantum computers proposed by DiVincenzo in 1995.

**Elementary Particle.** A particle that has substructures. There are many elementary such as photon, electron, proton and neutron.

**EDVAC (Electronic Discrete Variable Automatic Computer).** The Neumann-type computer completed at University of Pennsylvania in 1951.

**ENIAC (Electronic Numerical Integrator And Computer).** The first electronic computer which was developed at University of Pennsylvania, USA in 1946 by Eckert and Mauchly.

**Entangled State.** The state in which each of pairs of generated particle cannot be described independently. It cannot be described by tensor product.

**EPR Paradox.** The paradox, given by Einstein, Podolsky and Rosen in 1935, which demonstrated the difficulty of the early formulations of quantum theory.

**FANOUT.** Quantum gate.

**Flying Qubit.** A qubit used in quantum communication.

**Fredkin Gate.** Universal reversible quantum gate.

**Grover's Algorithm.** A quantum search algorithm proposed by Grover in 1996.

**Holevo Bound.** An upper bound on the amount of classical information that can be accessed from a quantum ensemble in which the information is encoded, which was proposed by Holevo in 1998.

**Matrix Mechanics.** A formulation of quantum mechanics using matrix proposed by Heisenberg in 1925.

**Measurement.** An action determining a particular property of a quantum system.

**Measurement Problem.** The problem of the connection between a quantum system and reality.

**Modern Physics.** Physics including quantum mechanics and Theory of Relativity.

**Neumann-Type Computer.** A computer which is based on the model proposed by von Neumann in 1945.

**NMR Quantum Computer.** A quantum computer based on NMR (Nuclear Magnetic Resonance).

**Observable.** Measurable physical value described by the hermitian operator.

**Pauli Gate.** The quantum gate based on Pauli matrices.

**Quantum Computer.** A computer based on a computational model which uses quantum mechanics.

**Quantum Computing.** The area related to quantum computers.

**Quantum Gate.** The model of quantum computers proposed by Deutsch in 1988.

**Quantum Logic.** The logical system for quantum mechanics based on Hilbert space, proposed by Birkhoff and von Neumann in 1936.

**Quantum Mechanics.** The subfield of physics which studies phenomena at the micro level.

**Quantum Number.** The number which can characterize various quantum states.

**Quantum Parallelism.** The method of computation in quantum computers performing several computations simultaneously.

**Quantum Physics.** Quantum theory based on quantum mechanics.

**Quantum Teleportation.** A quantum algorithm for teleportation which transmits qubits from one place to another.

**Quantum Turing Machine.** The computational model of quantum computers, which is a reformulation of classical Turing Machine, proposed by Deutsch in 1985.

**Quantum Walk.** A quantum algorithm for random walk.

**Qubit.** The basic unit of information in quantum computers.

**RSA Code.** The code proposed by Rivest, Shamir and Adelman in 1978.

**Reversible Computation.** The computation whose computational process is reversible. It satisfies the laws of thermodynamics.

**Schrödinger's Cat.** The thought experiment, given by Schrödinger in 1935, which shows the defect of the Copenhagen interpretation.

**Schrödinger Equation.** A basic equation of quantum mechanics proposed by Schrödinger in 1926.

**Shor's Algorithm.** The first quantum algorithm for factorization proposed by Shor in 1994.

**SWAP.** Quantum gate. It is also called EXCHANGE.

**Theory of Computation.** A theory which studies the notion of computability. They include Turing Machine, recursive function and $\lambda$-calculus.

**Trapped Ion Quantum Computer.** A quantum computer based on trapped ion.

**Turing Machine.** An abstract machine proposed by Turing in 1936. It is one the most famous theories of computation.

**Uncertainty Principle.** The principle, proposed by Heisenberg in 1927, which says that the position and momentum of a particle cannot be simultaneously measured with arbitrarily high precision.

**Wave Mechanics.** A formulation of quantum mechanics using wave function proposed by Schrödinger in 1926.

**Wave-Particle Duality.** The hypothesis that all particles have both wave-like and particle-like properties proposed by De Broglie in 1924.

# Index

Printed in the United States
By Bookmasters